But It's a Dry Cold!

But It's a Dry Cold!

Weathering the Canadian Prairies

Elaine Wheaton

Front cover image © Winnipeg *Free Press*, by Joe Bryksa, 5 April 1997.
Reprinted by permission.
Back cover image © John Perret/Light Line Photo.
Cover and interior design by John Luckhurst/GDL

We acknowledge the financial support of the Government of Canada through the Book Publishing Industry Development Program for our publishing activities.

The publisher gratefully acknowledges the support of the Department of Canadian Heritage and the Canada Council for the Arts for our publishing program.

Printed in Canada.

05 06 07 / 5 4

Canadian Cataloguing in Publication Data
Wheaton, E.E. (Elaine Ester)
But it's a dry cold

Includes bibliographical references and index.
ISBN 1-894004-01-9

1. Prairie Provinces — Climate — Miscellanea. I.Title.
QC985.5.P7W53 1998 551.65712'02 C98-910166-5

Fifth House Ltd.
A Fitzhenry & Whiteside Company
1511-1800 4 St. SW
Calgary, Alberta, Canada
T2S 2S5

1-800-387-9776
www.fitzhenry.ca

Contents

Acknowledgements

Authors in pursuit of their writing goals can be terrible creatures. They ignore mail, phone calls, laundry, and other duties. They incessantly demand data, publications, photos, information, answers to crazy questions, and some almost impossible tasks.

Many people were caught in my whirlwind of writing. I am thankful to the following colleagues at the Saskatchewan Research Council: Carol Beaulieu, Research Technologist, for research, discussions, graphics, technical editing, and many other tasks; Virginia Wittrock, Associate Research Scientist, for information, graphics, discussions, and technical editing; Brenda Tacik, Technical Communications Specialist, for technical editing; Julie Raven, Library Technician and Colleen MacLeod, Information Coordinator, for successfully hunting down references; and Leanne Crone, Administration Officer, for word processing.

Thanks to Fifth House Publishers for setting me on this quest and suggesting themes and to Charlene Dobmeier, Managing Editor, for many roles including editorial advice and photograph selection. I am grateful to Donald Ward, Ward Fitzgerald Editorial Design, for editing. Don tamed the technical jargon and smoothed the current of words. Many others, acknowledged in the book, provided information, photos, and answers to numerous questions.

My family is owed considerable appreciation for their patience and encouragement during this work. My husband, Dale Young, read and commented on drafts. My sons, Sean and Trent Young, offered feedback and often enquired how the work was going. My mother, Edna Wheaton, read portions of the manuscript and was very encouraging. Many of the weather anecdotes originated from my father, James Wheaton. Thanks, Dad, for sharing them.

My thanks to all of you. I hope you enjoyed your part in creating this book.

Climatologists are always compiling statistics and anecdotes about the weather, even when they're not writing books. If you have a weather adventure you'd like to share, please write to me at:

Saskatchewan Research Council
15 Innovation Blvd.
Saskatoon, SK S7N 2X8

Preface

What is it about weather that fascinates people? And why prairie people in particular? Our preoccupation—some might call it an obsession—with the weather must be perplexing to folk from milder climates who have to suffer through the boredom of consistent temperatures and conditions. (What *do* they talk about?) My own obsession is easily explained: I'm a climatologist. I'm *supposed* to find the weather fascinating. But that begs the question: what led me to be a climatologist in the first place? Well, I've always been the curious type, and what could be more curious and spectacular than the climate of the Canadian prairies?

The prairies is one of the most sensitive and vulnerable weather regions on Earth. This makes it an excellent place to study how weather and climate affect people and the environment, and vice versa. Growing up on a farm, I learned these lessons early. I remember seeing the despair on my parents' faces as three minutes of hail ruined an entire year's work. I remember their anguish as dust storms ripped the precious topsoil from the fields and sandblasted or buried new crops. I shared their elation when their labours were rewarded with beautiful crops and a bountiful harvest. And of course, the Great Depression has left its mark on everyone. We've all listened to tales of hardship and perseverance from men and women who fought an appalling struggle against the heat and the dust. You didn't have to live through the Depression to grow up with it.

Prairie people have a saying: "If you don't like the weather, wait five minutes." Winter and summer and all points between, the prairies have some of the most changeable and exotic weather in the world. We're captivated by it. We're sometimes *held captive* by it, too. Blizzards, floods, and storms rage across the prairies, wreaking havoc on the landscape and in our lives. We cannot ignore the weather. It entertains us. It humbles us. It can destroy our life's work, or make us rich.

But despite our frequent grumbling, we're proud of our weather—or rather, we're proud of our ability to endure anything nature dishes out. We routinely survive the extremes of almost every type of weather imaginable. And the hot spells become hotter, the cold spells longer, and the storms more magnificent with each telling. Bad weather aside, we live in a climate filled with splendour, with lush summer evenings and majestic and changeable skies, not to mention record amounts of sunshine in winter and summer alike.

My goal in this book is to improve our understanding of weather and climate, and to work toward harmonizing our activities with the conditions and consequences of our natural environment. Better use of meteorological and climatic information can lead to improvements in human health and safety, in agriculture and forestry, in water use and supply, in recreation and tourism, environmental protection and enhancement, and in countless other areas.

Prairie people have a great deal to gain, and at least as much to lose, from understanding or misunderstanding the weather. Our lives and livelihoods may well depend on being weather wise. Unfortunately, few books about prairie weather and climate have been written for the public; valuable and intriguing information remains locked up in scientific journals and textbooks that are inaccessible to the ordinary person. *But It's a Dry Cold* attempts to transfer some of this knowledge to those who live with and experience prairie weather in their everyday lives. As you will find, there is much to discover about our weather and climate.

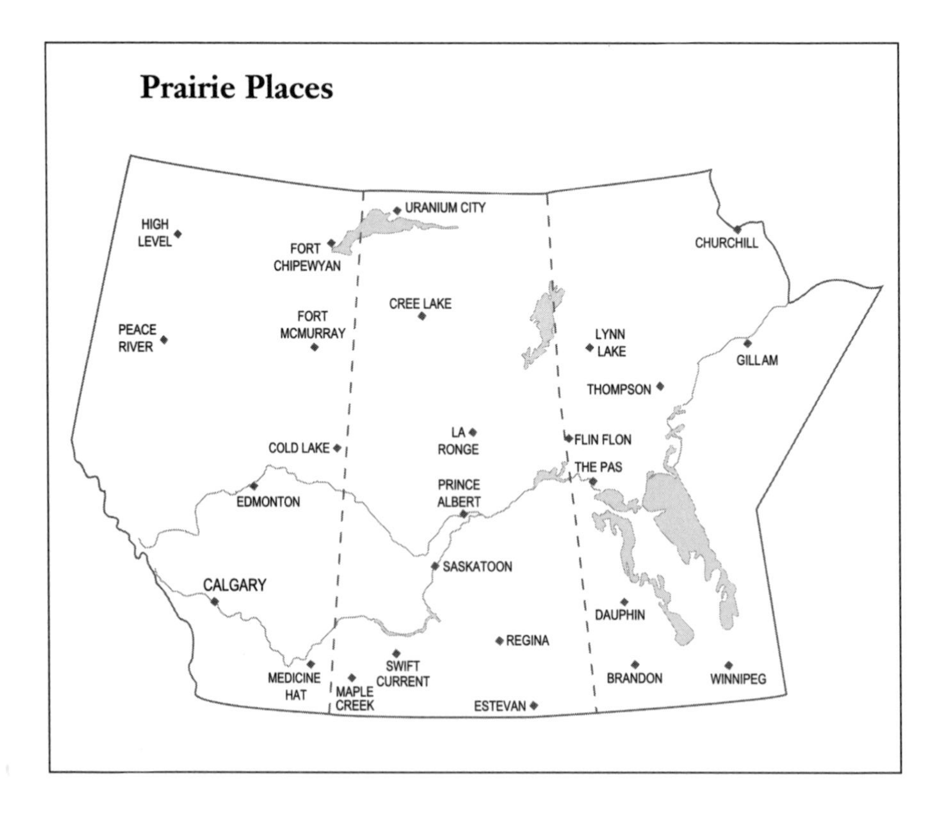

Chapter One

Climate Wars

In February, under a clear blue sky, a ground blizzard blowing on a north wind hid the Earth, and the whole world seemed one vast river of moving snow with only power posts and buildings showing that the ground was still there.

Andy Russell, *The Canadian Cowboy*

The Canadian prairie experiences every type of storm except hurricanes and tsunamis. Blizzards, thunder storms, dust storms, wind storms, hail storms, tornadoes—even the occasional mud storm—are all part of prairie life. But they still manage to surprise us.

Blizzards and Snowstorms

People in all regions of Canada are at risk from blizzards. Lives are lost not only by exposure to the elements, but indirectly from traffic accidents, carbon monoxide poisoning, falls, heart attacks from over-exertion, and house fires. These storms need to be treated with respect. So let's have a look at them. What are blizzards? Where and when do they strike? And which blizzards are the most severe?

Winter storms and excessive cold claim over 100 lives in Canada every year, more than the combined toll from hurricanes, tornadoes, floods, extreme heat, and lightning. Blizzards are easily the most feared and perilous of all winter storms.

Phillips 1993: 72

Not all storms with blowing snow make the grade as blizzards. Definitions vary, depending on where you live. The Alberta criterion for temperature, for example, is only 0°C. In general, though, a blizzard meets these conditions:

- temperature of -12°C or lower
- wind speed of 40 kilometres per hour or higher
- visibility of less than 1 kilometre
- duration of at least 4 hours

A Winter Ordeal

A few years ago, I was on my way to my son's hockey game in Saskatoon—we live ten minutes south of the city—when I heard an Environment Canada warning that a severe snowstorm was heading into Saskatchewan from Alberta. It seemed a pleasant enough Sunday afternoon for early December, though, and when I arrived at the arena people were scoffing at the idea of an imminent storm. But, being prairie people, we were used to expecting the unexpected, and most of us went through an informal checklist of storm preparations, including places to stay if we were stranded in the city.

By Monday, forecasters had pegged down arrival times: the storm would hit Saskatoon at about noon. Right on schedule, the blizzard howled into the province. Earlier that morning we had watched the barometric pressure take a steep dive, indicating an oncoming storm. By 11:00 AM the pressure needle had dipped to 923 kPa, almost the bottom of the chart. The temperature, a pleasant 0°C at 11:00 AM, began dropping rapidly. By midnight it was -21°C and still falling.

Other changes were also occurring. Visibility was nearing zero in places as the wind whipped up the snow, creating whiteouts. The wind speed, less than thirteen kilometres per hour at 11:00 AM on Monday, was over seventy-seven kilometres per hour by 3:00 PM. Wind speeds were even higher in open areas outside the city. Only three centimetres of new snow fell during the storm, but there was enough loose snow on the ground already to be picked up by the wind to make driving dangerous.

Driving home from work that day was terrifying. I often lost sight of the middle of the highway. My car was crawling along when I saw the flashing lights of an RCMP cruiser. The officer pulled me over and informed me that the highway was closed. I would have to return to the city. But I was halfway home already, I told her, and the road was more sheltered from this point. Reluctantly, she let me continue, so on I went. The school buses had delivered my sons home early, and I found them enjoying their storm holiday.

We were luckier than others. An eleven-car pile-up occurred just north of the city, and about seventy cars were trapped behind a semi-trailer that was blocking a nearby highway. Injuries were light, though; people had made it through one more prairie blizzard.

(Cameron Cardow, "Cam", Regina *Leader Post*, 3 January 1997)

A Future Prime Minister Nearly Freezes to Death

On 11 March 1908, young John Diefenbaker and his Uncle Ed decided to stay after school to attend a concert. Diefenbaker was wearing only leather shoes and a cloth coat, having discarded his heavier winter attire that morning owing to the relative warmth of the day. The concert began early in the evening. Within an hour a blizzard was blowing outside, and it was getting worse. Ed and John set out for home in their open sled. As the storm worsened, the temperature fell rapidly. At one point they caught sight of the lantern in the Diefenbaker farm yard some four kilometres away, but it was soon blocked out by the wind and blowing snow. Then the roadway was no longer visible. The horse veered away from the wind, then came to an abrupt halt by a willow bluff.

"We had lost all sense of direction," Diefenbaker wrote. "Uncle Ed, who had not discarded his winter wear and had on his felt boots, dog-skin coat, and Balaclava cap, gave me the only covers we had, two horse blankets. As the snow drifted high around the sled, affording some protection from the wind which screamed and moaned with an almost human sound, I fell asleep in the hay on the floor of the sled. Many times I was awakened by Ed during the night. Asleep, I dreamed of being in a warm room, then again of picking apples in my Grandfather Diefenbaker's orchard . . . Awake, I would shiver, my teeth chattering, my uncle's words of encouragement lost on my numbed senses. My feet and legs lost all feeling. At daybreak, the blizzard was still raging. Then, almost without warning, it subsided. Uncle Ed got the horse free from the shafts, and the three of us started through the drifts. My legs were like blocks of wood. After a long hour or more, we arrived at the homestead."

There, to his surprise, Diefenbaker learned that his parents had assumed they were storm-bound at the school, and had not been worried in the least. For years afterwards, the future prime minister was afraid to go outside, even to the wood pile, during a blizzard, and the eleventh of March was a day he remembered for the rest of his life.

from an account in *One Canada: The Crusading Years*
by John G. Diefenbaker

Treacherous travel (Richard Marjan, *Star Phoenix* Photo, 16 December 1996)

Some Severe Prairie Blizzards

Location and Date	Storm Characteristics	Storm Effects
Southern Prairies 24 March 1904	Fierce 3-day blizzard with 30 cm of snow, 100 km per hour winds and -18°C temperatures.	Five trains were snowbound between Winnipeg and Calgary.
Southern Prairies 1906–07	The blizzards started in mid-November, and a series of bad storms followed. One blizzard after another, and record low temperatures made this winter the measure of catastrophe.	It was the worst winter of the early 1900s. Cowboys fought to keep cattle and themselves alive in the southwestern prairie, but it was a losing battle. The spring melt uncovered literally herds of dead animals.
Regina, Saskatchewan 30 January 1947	The blizzard raged for ten days, forming snowdrifts up to 1 km long. More trouble was caused by the extreme cold following the storm. Temperatures dropped to -43°C by 1 February 1947. The storm also battered several other southern areas including Rosetown, Davidson, Weyburn, Swift Current.	One train was buried in a snowdrift 8 metres deep. Regina was isolated. For ten days all highways into Regina were blocked. Telephone lines were down. Supplies of fuel, food and livestock feed ran perilously low.
Southern Prairies 15 December 1964	Heavy snows, 90 km per hour winds, and -34°C temperatures. This was a "Great Blizzard."	Three people froze to death and thousands of animals perished.
Winnipeg, Manitoba 4 March 1966	36 cm of snow and 120 km per hour winds.	These conditions paralyzed the city for two days.
Regina and southern Saskatchewan 6 February 1978	Snowfall was minimal; winds and blowing snow were the culprits. Seven days of blowing snow and dangerous wind-chills were recorded. Winds peaked at 98 km per hour at Regina on 6 February.	Regina was snowbound for four days. Snowmobile teams were organized to search for stranded motorists. Several schools were closed, including those in Regina and Moose Jaw. The University of Regina closed both its campuses. Snowdrifts started to cover Regina homes, and many people had to be dug out. Many livestock perished.

continued on page 8

Location and Date	Storm Characteristics	Storm Effects
Southern Alberta 14 May 1986	The 2-day storm was described as the worst spring storm in Alberta. Snow was knee deep and the wind was 80 km per hour.	Dozens of communities were without services.
Winnipeg, Manitoba 8–9 November 1986	A major storm dumped 35 cm of snow on the city; winds gusted to 90 km per hour, bringing severe blowing snow and zero visibility across southern Manitoba.	Clean-up costs approached $3 million. The city did not get back to normal for a week. Traffic all across southern Manitoba was disrupted.
Southern Manitoba 5–6 April 1997	This was one of the great recent storms. Its swath of damage was about 350 km wide, extending north of Winnipeg and south into North Dakota. Records were set for snowfall (48.6 centimetres for the month) and duration (24 hours). Wind speeds gusted to 85 km per hour. The storm also brought zero visibility, freezing rain, and ice pellets.	Several people died of heart attacks while battling the storm. One boy was buried in his snow tunnel and asphyxiated. Highway accidents and strandings were common. Highways, schools, and airports were closed. Power outages were extensive throughout southern Manitoba. Estimated cost to taxpayers was $3 million.

(Phillips 1990; Phillips 1993; Department of Transport, ND; Environment Canada 1978; *The Leader-Post*, 31 January and 3 February 1947, 7 to 8 February 1978; Winnipeg *Free Press*, April 1997)

Southern Saskatchewan is the blizzard capital of Canada. On average, more than thirty hours of blizzard weather per year are clocked in places such as Swift Current and Regina. Most blizzards last an average of twelve hours, but extremes of forty-eight hours have been recorded.

Legendary Blizzards

Although most regions of Canada experience blizzards, the most legendary ones have taken place on the prairies. Several of the ten worst blizzards in Canadian history were prairie storms: Regina, 1947; Winnipeg, 1966; southern Alberta, 1986; and Winnipeg again in 1986 and 1997. Damage from blizzards includes lost lives—both human and animal—stalled and buried vehicles, collapsed buildings, power outages, and snow-clearance costs rising into the millions of dollars.

Perhaps the blizzards of forty and fifty years ago seem worse than the ones we experience today only because people were less able to cope with them. The great blizzard of 1947 in southern Saskatchewan, for example, crippled the city of Regina and many other, smaller communities. Telecommunications and rail and road transportation were completely shut down. Surrounding towns and villages were left to cope on their own.

Are we better able to cope with these hazards now? With modern technology, our infrastructure has improved considerably: underground telephone, power, and gas lines, improved roads and snow ploughs have made communication, heating, and transportation more reliable. We also have better clothes, better housing, and more effective and reliable methods of food storage. Cell phones enable us to call for help when stranded, and snowmobiles and four-wheel drive vehicles make travelling in deep snow much easier, even fun.

On the other hand, blizzards seem to be getting less frequent and intense. Snow cover amounts have decreased over time, giving the wind less material to swirl and pile up during a blizzard, and average winter temperatures have steadily increased since the turn of the century. The numbers of cold

Blizzard Precautions

- Winterize your home and vehicle before the cold weather arrives. Prepare emergency packs for your cars and ensure your home heating system is in good working order.
- Maintain a supply of heating fuel and food. If you've been warned early enough, stock up on food, especially things that require no preparation in case the power goes off.
- Ensure animals are properly sheltered, fed, and watered before the storm reaches its full force.
- Wait out the storm indoors. Blizzards can last for days at a time.
- If you must go out, tie a long rope firmly to the house and to the other buildings you visit to provide a way back. Keep a tight grip on the rope at all times.

- Be prepared in case of power failure. Check battery-powered equipment, candles, flashlights, and portable radios before the storm hits.
- Be alert for fire hazards from overheated stoves, fireplaces, heaters, or furnaces.
- Keep an adequate supply of prescription drugs on hand, in case you run out during the storm.
- Use caution shovelling snow after the storm. It is hard work for people who are in less than excellent physical condition, and can even bring on a heart attack.
- Pace your outdoor activity. Be alert for signs of frostbite. Avoid strenuous activity in extremely cold temperatures, as the heart must work much harder to pump blood through constricted vessels in arms and legs.

spells are also decreasing. So perhaps we only think we're coping better, when in fact we're fooled because the storms are fewer and weaker.

If the great blizzard of 1947 reappeared, would it push Regina as close to the edge as it did then? Probably not. But we cannot afford to feel complacent. Prairie towns and farms are disappearing, leaving fewer places for stranded motorists to seek help. Blizzards are still dangerous. People still die in them. We ignore their power at our peril.

Snowstorms in Summer

Summer snowstorms have to be the cruelest of weather, attacking us not only psychologically, but physically and economically as well. We're usually prepared for blizzards from September to April, even May. But when freak snowstorms strike in the summer, we get angry.

On 22 August 1992, people in southern and central Alberta woke up to a heavy blanket of wet snow. Only a few days before, the area had been sweltering under a 30°C hot spell. The storm arrived in time for the weekend. Vacationers had to replace their bathing suits and light jackets with boots and snow suits. Farmers, in the middle of harvest, watched as snow and ice flattened their crops. They had to improvise, using hay rakes to coax flattened stalks of grain upright, and putting other machinery to novel uses in order to lift and harvest their crops. Gardens were freeze-glazed, and the wet snow downed power lines and trees and created hazardous driving conditions.

Newspaper headlines reflected the disgust and dismay most people felt:

Early Winter Costly
(Regina *Leader Post)*

It's the 46 Year Blues as Snow Hits—
Arctic Front Plays Havoc with Gardeners, Farmers
(Calgary *Herald)*

Snow Buries Farmers' Hopes
(Calgary *Herald)*

Snow in August Illustrates Weird Weather on Prairies
(Saskatoon *Star Phoenix)*

If Only Politicians Were Like Weather—Every 46 years, whether we need it or not, it has snowed in August in Alberta. . . . If only Canada's constitutional problems could be put on a similar schedule.
(Calgary *Herald)*

Nature rules when it comes to farming. These men near Browning, Saskatchewan, 12 September 1903, must have been surveying the damage caused by an early snow fall. (South Saskatchewan Photo Museum)

An Arctic air mass was the culprit, bringing record low temperatures and rain that soon changed to snow along the cold front. The disturbance was widespread, covering a large part of Alberta and as far east as Swift Current in southwestern Saskatchewan. It also edged into northwest Saskatchewan, dropping as much as twenty-three centimetres of snow in the Meadow Lake area. An extended frost took a further toll on crops and farmers, who had been headed for a near-record harvest before the storm.

The snow ruined the weekend for vacationers, too. Hundreds of campers fled the campgrounds for the warmth of home. A man at Pincher Creek, Alberta, found himself trapped in his tent by a snow bank, and had to chew his way out through the flyscreen.

Summer snowstorms bring a special type of havoc to the prairies. Beside the usual deaths from exposure and traffic accidents that winter blizzards bring, summer blizzards also destroy crops, gardens, trees, annual vacations, and newborn birds and animals. Cattle that have been turned out to pasture will often bunch together for warmth, smothering those on the inside, particularly if the snow is heavy and wet. Physically or mentally, neither people nor animals are prepared for summer blizzards.

Friends and Fun in the Storm

There is a bright side to blizzards and snowstorms. They can improve or create friendships, as neighbours and strangers alike rely on one another to help out. Prairie people show their generosity by sharing their homes and food with those stranded in the storm.

My parents often provided a temporary home for children when our farmyard became the "end of the line" for the school bus because of bad weather. With the blizzard howling outside, birds and animals snuggling in the yard, several times the usual number of kids would sit around our supper table. My mother and father had endured storms that had left them snowed in for a week or more, so they always had enough food stocked up to outlast the worst that winter had to offer.

Blizzards have also been known to provide forced respites from busy schedules. When visibility is close to zero and the yard is piled high with snow, it is not wise to venture out to work or school. These unplanned days often seem like more of a holiday than a forced day off.

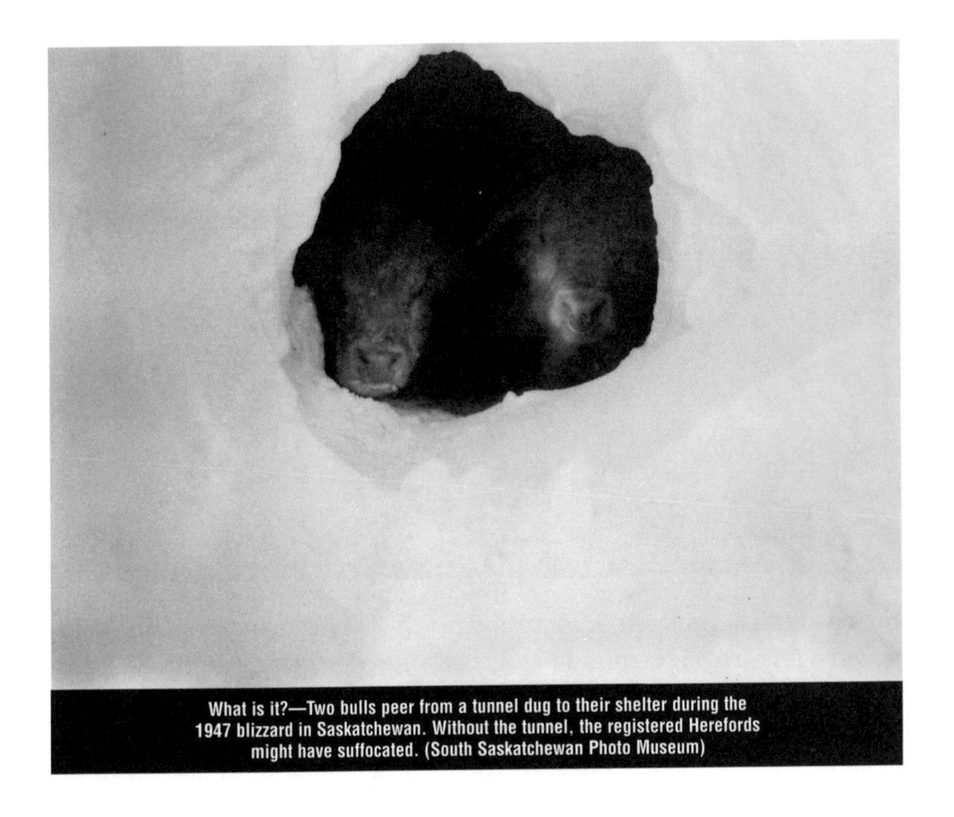

What is it?—Two bulls peer from a tunnel dug to their shelter during the 1947 blizzard in Saskatchewan. Without the tunnel, the registered Herefords might have suffocated. (South Saskatchewan Photo Museum)

Dry, Dirty, and Dangerous

And, oh, the dust clouds. How I remember them. Brown ones, red ones, yellow ones, made from the soil of thousands of farms across the prairies. One big dust cloud blocked out the sun for days. As it moved across the country, it covered the land in darkness. We had to keep the lanterns lighted all day. Some people in the cities thought the end of the world had come.

David Booth, *The Dust Bowl*

Most natural disasters are linked to extreme weather conditions and events. Drought, one of the most damaging of environmental phenomena, is invariably accompanied by severe dust storms, or "black blizzards." A dust storm is defined internationally as an atmospheric disturbance with moderate to strong winds (twenty to forty-nine kilometres per hour), blowing soil particles, and visibility reduced to one kilometre or less at eye level.

Climate and weather continue to be the chief factors affecting agriculture, and drought and its accompanying dust storms are among its chief

A wall of dust during the 1930s, Alberta. (Glenbow Archives/NA-2496-1)

threats. Dust storms can cause serious environmental and economic damage, ranging from soil erosion and vegetation destruction to traffic fatalities.

Dust storms are visible indicators of widespread, severe, and costly wind erosion. One dust storm can severely reduce the long-term fertility of the soil, its water-holding capacity, and its ability to withstand further erosion. The on-farm costs of wind erosion in the prairie provinces are estimated at $249 million annually—sufficient to send the entire populations of Lethbridge and Red Deer, Alberta, to Hawaii for two weeks every year.

Dust storms can cause anything from extra cleaning problems to traffic

Two Die in Highway Accident

Except for the cause of this accident—a dust storm—this event would not be unusual. Six vehicles were involved—three semi-trailers and three cars. Winds in the Kindersley-Saskatoon area were gusting up to 100 kilometres per hour, and visibility was often at zero. This type of accident would be rare anywhere in Canada except on the prairies, where dust storms are often extremely hazardous. Anyone who has experienced such an event would not have to be convinced of this.

Saskatoon *Star Phoenix*, 1 June 1984

disruptions and fatalities. They bury ditches and roads, crops and wetlands, fences and trees; they sandblast buildings and cars. The off-farm costs of wind erosion, though more difficult to quantify, may be as high or higher than the agricultural costs.

Dust storms can occur at any time, even during periods of flood. In 1988 in Manitoba, one team of water resources personnel had to drive through a dust storm to get to the flood zone they were monitoring. Dust storms

Sand dunes begin to bury the northern forest, south shore of Lake Athabasca. (Zoheir Abouguendia)

This ditch collected soil—not snow—as intended. (Elaine Wheaton)

can also uncover items long lost under layers of soil, including things of archaeological value. (On the other hand, they are also responsible for burying many of these items in the first place!)

Breathing Dirt

Wind-blown dust is a major source of particulate air pollution (that is, pollution consisting of small, separate particles) on the prairies, especially in Saskatchewan. The health of people and animals alike can be seriously affected by these storms, which can cause or contribute to eye, lung, and allergy problems. Even at moderate concentrations, air-borne dust has been linked to increased sickness and death.

Missing and Dangerous

The drying and blowing of salt lake beds during drought years makes life difficult for people with health problems. In 1988, for example, Old Wives Lake, near Moose Jaw, Saskatchewan, disappeared; it dried up and started to blow away. So many people were bothered by the salty dust in the air that the Centre for Agricultural Medicine at the University of Saskatchewan undertook a study of the air quality and its health implications.

Old Wives Lake, with an area of 177 square kilometres, is the fourth-largest alkaline lake in North America—when it has water, that is. Massive amounts of dust, composed mostly of sodium sulphate, were produced during the drought years of 1987 and 1988. The salt was deposited in a plume extending fifteen kilometres from the southeastern end of the lake, ranging in width from ten kilometres at the lake to five at the tip of the plume.

Area residents reported increases in nasal, throat, and eye irritation, and numerous respiratory problems. Livestock, too, suffered weight loss and nasal and eye irritation. Little is known about the health effects of inhaled sodium, magnesium, and calcium, and by the time of the study, rainfall had soaked the area, destroying any chances for further research.

Wildlife also suffer during dust storms. At Sherlock Lake, a small saline body of water in western Saskatchewan, 300 dead or dying geese were found to be suffering from salt crusting their feathers, severe diffuse pulmonary edema (fluid in the lungs), cardiac dilation (enlarged hearts), and other problems. These were very sick animals, and their condition reminds us that we all need to take precautions during dust storms. Responsible soil and water conservation practices help minimize conditions that lead to dust storms in the first place. If it's too late for that, stay inside and avoid travelling.

Some Physical and Economic Effects of Wind Erosion and Deposition

Physical Effects	Economic Consequences
Soil Damage	**Soil Damage**
(1) Fine material, including organic matter, blown away (2) Soil structures degraded. (3) Fertilizers and herbicides lost or redistributed.	(1-2) Long-term loss of fertility, resulting in lower returns. (3) Replacement costs of fertilizers and herbicides.
Crop Damage	**Crop Damage**
(1) Crops buried by deposited material. (2) Sandblasting cuts down plants or damages foliage. (3) Seeds and seedlings blown away. (4) Fertilizer, herbicides, and pesticides redistributed, with resultant losses from original locations and hazards created in others. (5) Soil-borne disease, weed seeds, and pests spread to other fields.	(1-5) Yield losses give lower returns. (1-3) Replacement costs and yield losses due to lost growing season. (4-5) Increased herbicide and pesticide cost.
Other Damage	**Other Damage**
(1) Soil deposited in ditches, hedges, along fences, on roads, in reservoirs, lakes, and streams. (2) Fine material deposited in houses and on vehicles. Paint and glass on farm machinery abraded, machinery "clogged." (3) Farm work held up. (4) Decreased visibility results in transportation and communication interruptions, accidents. (5) Air pollution. (6) Changes in the Earth-atmosphere energy budget, such as increases in atmospheric cooling.	(1) Costs of removal and redistribution of sediment. (2) Cleaning and repainting costs. (3) Loss of working hours; productivity declines. (4) Decrease in transportation and communication efficiency; costs of accidents and fatalities. (5) Adverse effects on human health from dust inhalation; costs of health care. Pollution of water bodies; costs of water treatment. Environment degradation from air pollution.

Adapted from Wheaton (1992)

People out walking in and around Saskatoon on the evening of 10 June 1990 expected perhaps a bit of rain. But they returned looking as if they had visited a mud spa. A dust storm had swirled into town, with a rain storm on its heels. The result was mud rain as seen on the trunk of this car. (Elaine Wheaton)

Devils in the Dust

Dust devils—also known as dancing devils, dervishes, and sand augers—are small but vigorous whirlwinds that often last only seconds and pick up a spiralling column of dust, sand, and debris. They range in diameters from about three metres to more than thirty, with an average height of about 200 metres, though a few have been observed several thousand metres high. They can rotate both clockwise and counterclockwise, and commonly develop in dry regions on hot, calm afternoons with clear skies. If you're outside on such an afternoon and hear a moderate roar accompanied by a whirling pile of dust and debris skipping along an erratic path, you've met the prairie whirlwind. Hang onto your hat and have a good look because it will soon disappear.

The Macklin Dust Storm

The Saskatoon *Star Phoenix* ran a series of amazing photographs of a dust storm in the spring of 1990. I wrote the photographer, Barbara Kloster, to beg copies. She sent not only the photos, but a wonderful description of the storm:

> *The dust storm lasted for six days, but some days the wind was stronger than others. It started on May 23, at which time only certain fields were moving. On the 24th everything was moving, but the winds began to ease off by about six.*

Later that evening we were out for a walk and we noticed the winds had increased again.

By midnight you could scarcely see the street lights and it blew all day on the 25th. By 4:00 PM, the news media were reporting the worst dust storm in years. That evening it calmed down, and we had high hopes the storm had blown itself out.

However, on Saturday the winds were strong again, enough to keep the soil sifting for most of the day and night; this continued through Sunday and on Monday we had another day of ferocious wind. The skies were again black, but finally on the eve of the 28th, the winds died down completely.

From the Dirty Thirties to the Dirty Eighties

Early in 1935, a region in the south-central United States, including parts of Colorado, Kansas, New Mexico, Texas, and Oklahoma, was labelled "the dust bowl." The long period of drought coupled with soil loosening by land use, sparse vegetation, and strong winds resulted in severe dust storms throughout the area. The drought also created a region of severe, long-lasting, and frequent dust storms on the Canadian prairies. Farm abandonment, fuelled by drought, wind erosion, and crop destruction, was common throughout the decade. In 1937, the Saskatchewan Wheat Pool received 157 reports of serious crop damage due to wind erosion.

Highway 31, south-east of Macklin, Saskatchewan, obliterated by a dust storm. Some of the soil was hauled back onto the fields. (Barbara Kloster)

Unfortunately, these conditions are not relics of the past. The droughts of the 1980s—more frequent and severe than any since the thirties—resulted in many dust storms, even some in winter. The most severe drought occurred in 1988. Most of the dust storms that year occurred in west-central agricultural Saskatchewan. Swift Current suffered sixteen dust storms that year, while Winnipeg had the second highest total with eight. New crops are often blown out of the ground or buried during dust storms. In the spring of 1980, many farmers had to reseed their crops—an expensive, time-consuming, and frustrating delay that put the whole harvest at risk.

Isolines, Isotherms, and Isohyets

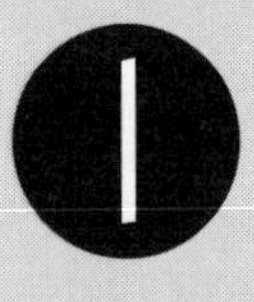

Isolines are lines joining places of equal value on a map. There are many types of isolines, including isotherms for temperature and isohyets for precipitation. Isotherms are lines joining points with equal temperature. Isohyets are lines joining places with equal precipitation.

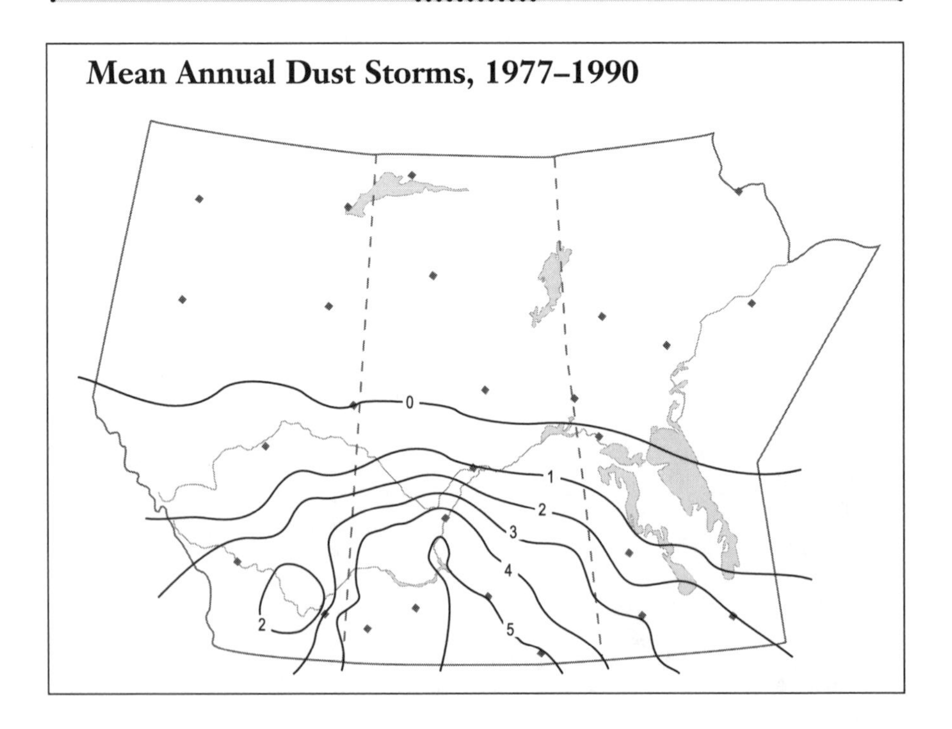

Mean Annual Dust Storms, 1977–1990

The Dust Storm Capital of Canada

South-central Saskatchewan is the centre of Canada's dust bowl, reaching up from the United States, past Moose Jaw and Regina, and northward to circle Saskatoon. This Canadian core usually has over four dust storms per year—as many or more, on average, than the desert areas of the midwestern United States. The difference is that in Canada the dust-storm season doesn't extend into winter (at least, not usually), whereas the season is year round in Arizona. The Canadian dust-storm period is much more concentrated.

The dust-storm risk decreases as you move north because of the slower wind speeds characteristic of forested areas. The risk also decreases rapidly as you move west into Alberta, less rapidly as you move east into Manitoba. But few places on the prairies are free of risk from these storms, and even Ontario and Quebec have had dust storms in drought years.

The number of storms to hit a given area can vary widely, from zero in one year to extremes of almost twenty in others. Regina holds the record for a single year, suffering nineteen dust storms in 1981. Moose Jaw comes second with fourteen, also recorded in 1981. Portage La Prairie, Swift Current, and Yorkton tie for third that year, with highs of twelve storms each. Alberta stations weren't in the same league. There was a high of eight dust-storm days recorded at Medicine Hat. But the only stations that recorded no dust storms that year were close to or within the boreal forest.

Peak dust storm years were 1977, 1981, 1984, and 1988. The lowest number occurred in 1983, only two years after the worst year, 1981. Since 1990 the numbers have been considerably lower.

How Long Do They Last?

We know how long dust storms last on Mars—about six months. We know somewhat less about their duration in Canada. As an object of serious study, our dust storms have been neglected by scientists, and the gaps in our knowledge of these amazing disturbances are huge. The neglect is certainly not due to a shortage of storms.

Dust storms on the great plains of the United States last almost seven hours, on average. The only year for which we have comprehensive analyses of Canadian dust storms is 1988, an extremely dusty year. The storms lasted an average of five hours in Manitoba and three in Saskatchewan. Although Manitoba generally had longer-lasting dust storms, the record for 1988 was set in Moose Jaw, Saskatchewan, where a dust storm raged for ten hours.

Because of the sparse network of observation stations, severe storms of all types can sneak through the prairies and remain virtually undetected by anyone not in the storm's path. So the numbers of severe storms have been underestimated.

Why Do They Occur?

Sometimes it seems as if the climate, topography, vegetation, soil conditions, and pattern of land use on the Canadian prairies were designed to produce dust storms. Drought is an ideal partner for these conditions, as weak and dying vegetation can no longer protect the soil. As soil particles dry, it is easier for the wind to pick them up.

Most dust storms occur in the spring, mainly April. There are at least ten to twenty times more in April than in January and February. A weaker, secondary peak occurs in late summer, usually August. This seasonal pattern is common to all the prairie provinces, but in Alberta the spring peak is much lower, and the secondary peak occurs much later—in October rather than August.

Although the Dirty Thirties are legendary, the "dirty eighties" remind us that dust storms go hand-in-hand with drought. But while they are more frequent during drought years, they can also occur during brief dry conditions, or even during a rainfall if the winds are strong enough. Even moist, healthy soil may begin to erode in a strong wind. On the other hand, if vegetation is well established and provides good cover, even a severe wind may not result in dust storms.

Dust storms in winter are less common because of the protective snow cover, but when snow is lacking, dust can fly. During the 1988 drought, for instance, the snow cover was short-lived and sparse, resulting in an expanded dust season. Most of southern Manitoba and Saskatchewan experienced black blizzards that winter.

Trends are masked by variations from year to year, but there appears to be a downward trend in dust storms since 1988, and perhaps a longer-term reduction since 1981. The possible reasons for this downward trend include above-normal rainfall for several summers and in several areas as well as the increasing use of conservation tillage practices in the prairie region.

Global Warming and Dust Storms

If Earth continues to warm, the numbers of dust storms will likely increase (see Chapter Four). The evidence already suggests that droughts will increase in frequency and severity, especially on the prairies. Dust storms will not be far behind.

The dust-storm core is expected to shift northeastward as the climatic zones move north, bringing higher risks of dust storms from the United States into Canada and putting our northern agricultural areas at risk. A shorter season of snow cover is also anticipated. This means that the dust-storm season could expand, as it did in extreme years such as 1988.

Dust in the air affects the atmosphere's heating machine, both scattering incoming sunshine and absorbing heat from the earth. An excess of dust low in the atmosphere may heat the earth, but only for a short time, as dust soon settles. On the other hand, dust in the upper atmosphere likely has a cooling effect. Recent work has emphasized the importance of dust in the process of climatic change. Scientists warn that we had better pay more attention to dust from deserts and agricultural land as dust particles may counteract some of the greenhouse effect.

Where Does All that Dust Go?

A single storm can transport tonnes of dust. This can translate into many truckloads of valuable soil. The sandier soil is dumped in ditches and shelter belts, and onto neighbouring fields. The finer particles become long-distance travellers, moving into the next provinces or even to the Atlantic Ocean. Measurements of dust concentrations and amounts of dust carried in dust storms are rarely made, though, so we really have no clear idea how much soil dust storms steal or how much they put in the air. But soil still blows in the dust bowl, and this could increase with global warming. Preparation, adaptation, and an adequate knowledge of the risks can help us avoid the damage wrought both by droughts and by their dark companions, the dust storms.

Tornadoes—Power and Havoc

Anyone who has experienced, or witnessed, the destructive power of a tornado, cannot fail to be moved by its awesome power. This freak of nature is probably the quintessential hazard . . . it strikes quickly, randomly and often without warning, leaving chaos and death in its wake.

Etkin and Maarouf 1995

A Monstrous Vandal

Animals battered and bruised, crops and gardens torn up or flattened, shingles ground to powder, windows shattered: what monstrous vandal could do such damage and still remain at large? A tornado, of course! This violently rotating column of air hanging from a cumulo-nimbus storm cloud is one of our worst weather thugs, and certainly the most destructive of all atmospheric phenomena on a local scale. It is nearly always seen as a funnel or tuba-shaped cloud. It can have a vortex hundreds of metres wide, with wind speeds estimated at 160 to more than 480 kilometres per hour. In the Northern Hemisphere it usually spins counter-clockwise, and its direction of travel is determined by the path of its parent cloud, which may travel along at up to forty-five kilometres per hour.

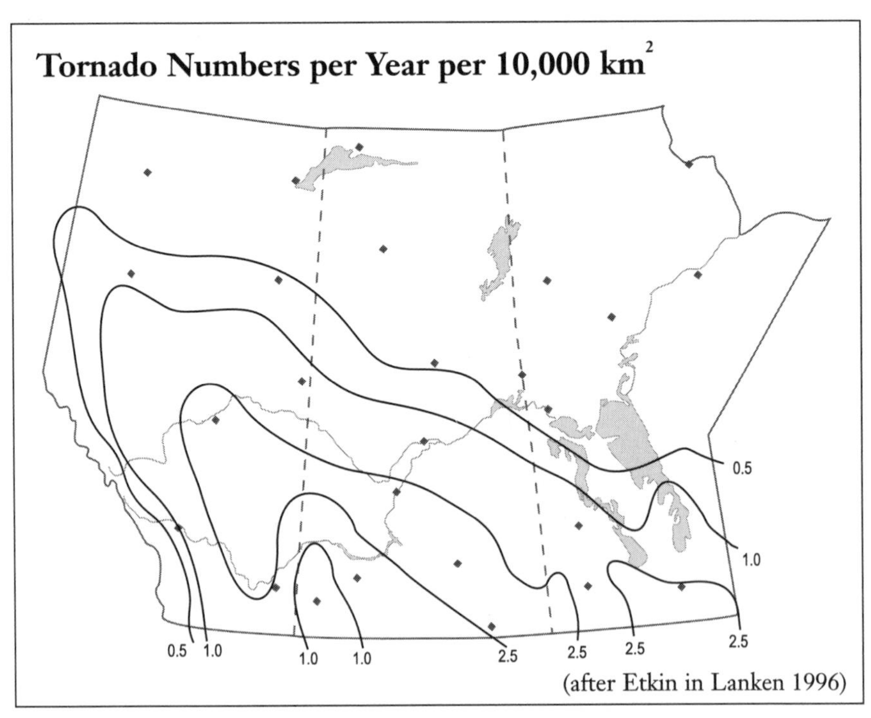

(after Etkin in Lanken 1996)

Although tornadoes occur fairly frequently on the Canadian prairies, they generally attack small areas, and the risk of damage at any given place is low. Still, they are nature's most locally destructive storm, and people need to be aware of the risks.

The Tornado Belts

Worldwide, tornadoes are most common in the United States, which averages 140 to 150 twisters a year. They can occur at any time of year or day, but are most common in the spring, and in middle-to-late afternoon.

Canada experiences the second-highest number of tornadoes in the world. This is a relatively new finding. At one time, tornadoes in Canada were thought to be rare, but their numbers had escaped notice because of our sparse population. Twisters often swept through the countryside undetected. No province in Canada is entirely free of these weather nightmares, but the areas of highest risk include southwestern and northwestern Ontario, southwestern Manitoba, and southern Saskatchewan. Areas of

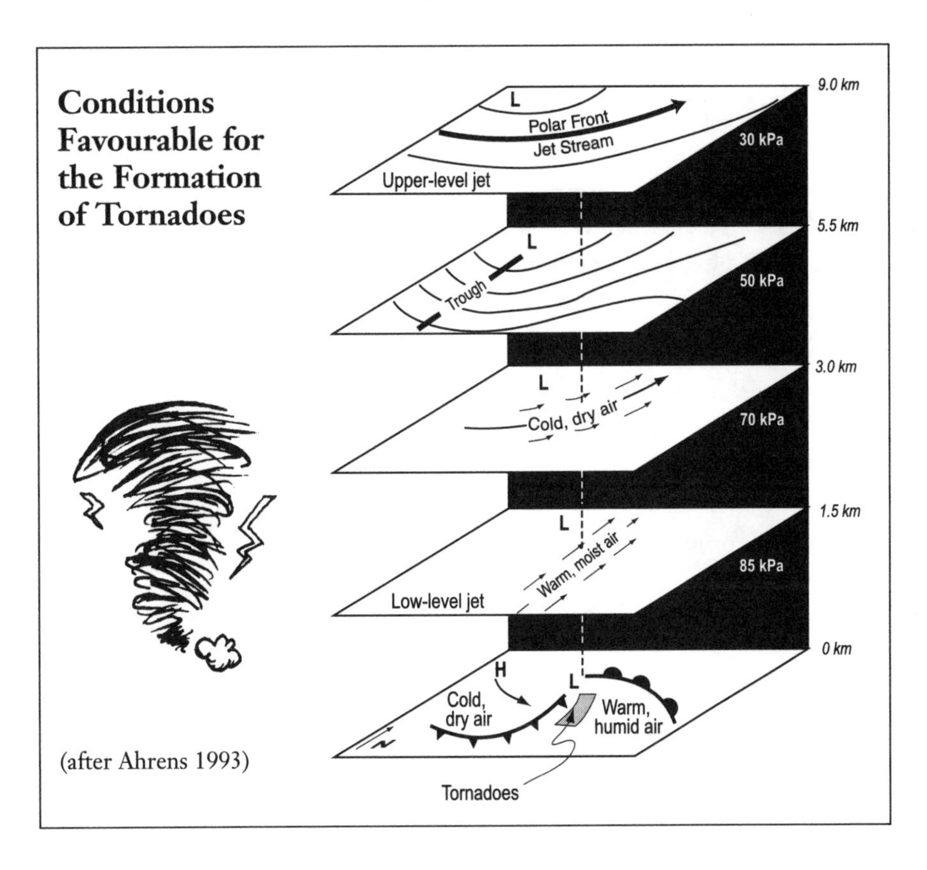

(after Ahrens 1993)

The 1912 tornado reduced many of Regina's buildings to toothpicks. The tornado remains at the top of the killer tornado list in Canada. (Saskatchewan Archives Board)

moderately high risk include western Quebec, northeastern Ontario, and south-central Alberta.

The greatest tornado risk is in the extreme southwest of Ontario, which averages three tornadoes a year over 10,000 square kilometres. The area of second-greatest risk is southwestern Manitoba and southeastern Saskatchewan near the international border. This belt averages more than two tornadoes a year per 10,000 square kilometres.

The exact physics of the origin of tornadoes remains a mystery, mostly because of their internal conditions: the winds are much stronger and the

The Six Worst Canadian Tornadoes by Death Toll

1) Regina "Cyclone," 30 June 1912. Twenty-eight dead and hundreds injured; $4 million damage.
2) Edmonton, 31 July 1987. Twenty-seven dead, 300 injured, 400 left homeless, $300 million damage.
3) Windsor to Tecumseh (Ontario), 17 June 1946. Seventeen dead, hundreds injured; damage conservatively estimated at $1.5 million.
4) Hopeville to Barrie (Ontario), 31 May 1985. Twelve dead and 155 injured; more than 1,000 buildings damaged and a total of $100 million damage.
5) Lancaster Township (Ontario) to St-Zotique to Valleyfield (Quebec), 16 August 1888. Nine (possibly eleven) dead and fourteen injured; extensive property damage.
6) Windsor (Ontario), 3 April 1974. Nine dead, thirty injured; $500 thousand damage.

(after Etkin and Maarouf 1995)

pressure much lower than can be measured by standard instruments. Even with specially designed instruments, one would have to predict the exact path of a tornado in order to place the instruments close enough to measure anything.

In general, a combination of events spawns severe thunderstorms with tornadoes:

- the collision of cold, dry air moving behind a speeding cold front with a warm, humid air mass behind a warm front
- an upper level trough of low pressure
- a polar front jet stream swinging over the region

Record Holders

The Regina tornado of 1912 (popularly but inaccurately known as the Regina "Cyclone") still tops the list as the most devastating of Canadian tornadoes in terms of fatalities; it killed twenty-eight people and injured hundreds more. The more recent Edmonton tornado of July 1987 killed twenty-seven, injured 300, and left hundreds more homeless. In purely financial terms, the Edmonton twister caused the greatest single storm loss in Canadian history, with damages of $300 million.

Tornadoes not only bring terrible human suffering; they also disrupt transportation and communication and inflict heavy economic losses.

The Rocky Mountain House "Cyclone"

8 July 1927, 2:30 PM

The tornado raged into Rocky Mountain House from the southwest. It first descended on the Atlas Lumber Yard, grabbing thousands of feet of lumber and hurling the boards around town like a deck of cards. The wood that didn't scatter was left in a tangle all over the lumber yard. Then the storm zeroed in on the west side of Main Street, demolishing a brick garage under construction and ripping off part of the roof of the Melton Hotel. It sliced off the second-storey front wall of the Mountaineer Shop next door and deposited it in one piece on its own doorstep. It skipped one building, then tore out two walls of Whitten's General Store, dropping one beside the store and slamming the other against the buildings across the street. The stock on the store shelves was left untouched.

The storm reduced McDermott's Hardware, the finest brick building in town, to a pile of rubble. Its assault on McLaren's Hardware was oddly selective: first, it ripped off the roof and tossed it aside, then it grabbed a bundle of pitchforks and hurled them over the town to land intact a mile away in a ditch by the fair grounds.

Meanwhile, the cyclone had snatched off the second storey of the blacksmith shop and collapsed the neighbouring livery barn, knocking out the attendant, George Wrigglesworth. Half a block north, it crushed the Wrigglesworth home, leaving the yard bare and scattering the contents of the house across town. Farther up the street, Teddy Brett was trying to put his car in the garage when the entire building was suddenly swept away. To the east, the storm flattened a hundred trees in the school yard and drove straws lengthwise into the wooden steps of the school. Nearby, Mrs. Good watched her dishpan circle the house before it shot over the school to wrap itself around a fence post at the fair grounds. On First Street West, the cyclone hit the creamery and the icehouse, smashed Strong and Parson's grain storage facilities, inflicted serious damage on several houses, banged a couple of cars around, slammed a piece of sidewalk into Edward's Garage and, perhaps in a moment of repentance, deposited a lean-to of unknown ownership on Ed Sim's property on the far west side of town.

Then, satisfied that every building in town was damaged to some extent, it tore off in a zigzag path northeastward, uprooting trees, sucking spouts of water a hundred feet high out of Gull Lake, destroying buildings, gardens, and crops, and, in one final stroke, tumbling a granary across a field near Wetaskiwin, killing the inhabitants—three newly landed immigrants—and so leaving three widows and ten fatherless children in Poland.

At 6:30 PM, torrents of rain with some hail swept into Rocky Mountain House to complete the damage, turning town and countryside into a quagmire. When all was finally calm, a blood-red sunset finished the day.

Although the town was left with damages running to a quarter of a million dollars—an almost unimaginable figure in 1927—the people could honestly be thankful for the timing of the storm. One week earlier, on the first of July, the whole town had been out celebrating the sixtieth anniversary of Confederation on Main Street. Had the storm struck then, the cost would have been measured in human lives instead of dollars.

based on an account by Esme James in *The Days Before Yesterday*

Even the weakest tornado can uproot trees, tip over granaries, wreck signs, and destroy chimneys. The strongest can wreak total devastation, crushing steel buildings, flattening structures of brick and stone, and picking up heavy machinery and tossing it through the air. Their only fortunate characteristic is that the destruction is localized along narrow paths and generally short distances.

The F-Scale

Tornadoes are classified according to the Fujita, or F-scale, which was developed for use in the midwestern United States. Considering environmental differences and variations in the types and variety of buildings on the Canadian plains, the F-scale was subsequently modified by tornado expert Dr. Alec Paul of the University of Regina for use in the Saskatchewan Tornado Project.

Modified Fujita Damage Scale

F0: Gale tornado (25–45 kilometres per hour)
Light damage, damage to chimneys, tree branches broken, small trees pushed over, empty granaries moved, phone lines knocked down, damage to sign boards.

F1: Moderate tornado (46–70 kilometres per hour)
Surface of roofs peeled off, quonsets blown over, curling rinks destroyed, farm machinery and implements destroyed, windows blown out, full granaries lifted and destroyed, fair buildings (grandstands) destroyed, small farm buildings destroyed, mobile homes moved or turned over, moving cars pushed off roads.

F2: Significant tornado (71–98 kilometres per hour)
Weak structures destroyed, barns destroyed, roofs torn off, boxcars knocked over, large trees uprooted, small objects (grass, straw) become missiles, chickens plucked, structures and livestock picked up and thrown, houses exploded, farm machinery scattered and destroyed, mobile homes demolished.

F3: Severe tornado (99–129 kilometres per hour)
Heavy cars lifted and thrown, combines picked up and thrown hundreds of metres, farm houses lifted and dropped twenty metres away, roofs thrown hundreds of metres, roofs torn from well-constructed houses.

F4: Devastating tornado (130–163 kilometres per hour)
General destruction (e.g., Kamsack 1944—four hundred buildings destroyed; Edmonton 1987—$300 million damage), well-constructed houses destroyed, large flying objects become missiles.

(after Paul 1995)

Tornado Trends

In recent years, there has been an increase in tornado sightings in both the United States and Canada, but it is unclear if this represents an actual increase in the number of tornadoes, or merely an increase in the number of people to see and report them. In addition to a growing population, the increase in sightings could stem from elevated public and media awareness, not to mention a growing fascination with the weather.

Tornado trends are difficult to assess. Cool summers generally produce fewer twisters; hot summers tend to produce more. Global warming may bring more hot summers, especially to mid-continental areas such as the prairies, and shifting climate zones will expand severe storm areas from the United States into Canada. Tornado experts caution us to expect a slight increase.

The Saskatchewan Tornado Project

Saskatchewan has documented more tornadoes than any other province, thanks to the Saskatchewan Tornado Project and Dr. Alec Paul. The Project has mapped the distribution of tornado occurrences from across the province for the period 1906 to 1991. The southeastern part of the province clearly experiences more tornadoes than the rest of the province,

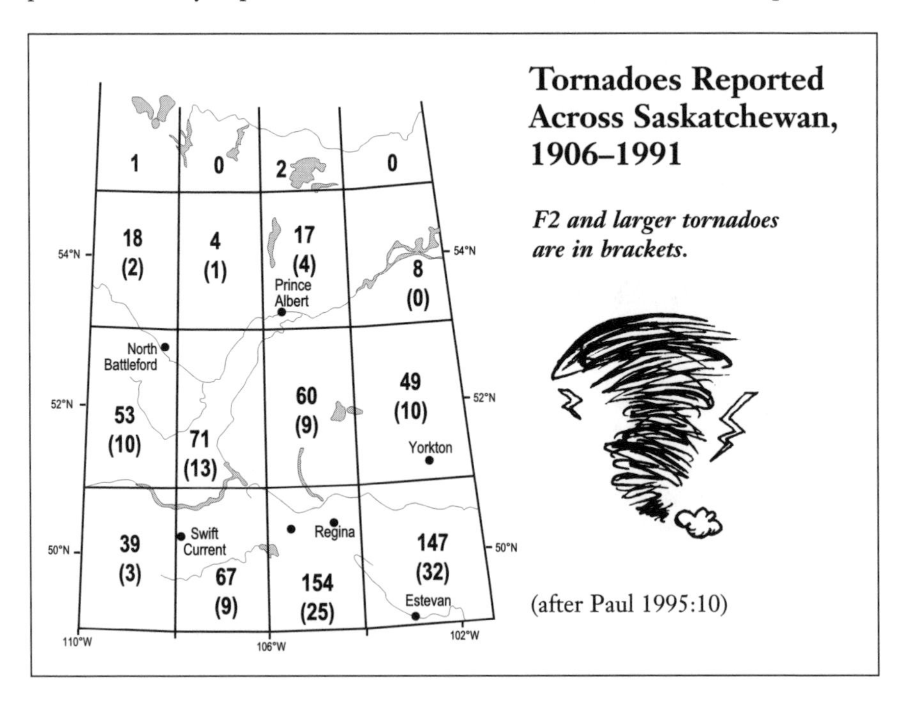

Tornadoes Reported Across Saskatchewan, 1906–1991

F2 and larger tornadoes are in brackets.

(after Paul 1995:10)

Fatal Tornadoes in Saskatchewan

Year	Date	Nearest Settlement	Fatalities	Persons Injured
1898	20 June	Benson	1	Unknown
1900	28 August	Wapella	5(?)	2
1907	8 August	Last Mountain Lake	1	7
1907	8 August	Zealandia	1	Unknown
1908	28 July	Fillmore	1	7
1909	1 July	Storthoaks (Ste. Antoine)	4+	28
1910	23 June	Palmer	3	8
1910	27 June	Weyburn	1	1
1910	3 July	Grandora	1	1
1912	30 June	Regina	28	80+
1913	14 August	Ogema	2	Unknown
1916	28 August	Atwater	1	4+
1919	27 June	Quill Lake	2	2
1919	27 June	Lanigan	1	Unknown
1920	22 July	Frobisher, Alameda	3	20
1920	22 July	Benson, Lampman	2	9
1923	16 June	Sceptre	1	Unknown
1923	16 June	Rosetown	1	3
1923	7 July	McGee	1	4
1924	July	Constance	1	Unknown
1926	14 July	Waldron	2	5
1927	18 June	Mozart	1	6
1932	31 May	Aberdeen	1	Unknown
1933	17 June	Saskatoon	1	Unknown
1935	1 July	Benson	1	2
1935	6 July	Smiley	2	Unknown
1935	28 July	Ile a la Crosse	1	3+
1944	1 July	Lebret	4	Unknown
1944	9 August	Kamsack	3	42
1963	29 June	Spy Hill	1	2
1976	3 June	Davidson	1	15
1979	10 July	Glasnevin	1	Unknown

(after Paul 1995)

with a high of 154 for the Regina block. But, surprisingly, there are reports of at least three tornadoes north of 55° north latitude (i.e., north of La Ronge). In all, 690 Saskatchewan tornadoes were recorded between 1906 and 1991. Seventeen per cent, or 118 of these tornadoes, were considered "strong"—that is, capable of causing significant damage. Are we likely to see another Regina "Cyclone"? Dr. Paul tells us that such an event is rare, recurring perhaps every 600 years.

Regina, at approximately one every 200 years, has the highest risk of experiencing an F2 or greater tornado, followed by Saskatoon at about one every 400 years, and Moose Jaw at one every 900 years. The lowest risk for agricultural Saskatchewan is in the northwest at the Battlefords.

Alberta Tornadoes

Dr. Keith Hage has been studying prairie tornadoes in Alberta for years. He lists thirteen fatal tornadoes to 1984. His work is based on data that are unconventional for the atmospheric sciences: archives, newspapers, and family and community histories.

Fatal Tornadoes in Alberta

Year	Date	Nearest Settlement	Fatalities	Persons Injured
1879	16 May	Saddle Lake	1	Unknown
1907	14 August	Wainwright	3	Unknown
1910	July	Sibbald	1	Unknown
1912	June	Oyen	1	Unknown
1915	25 June	Grassy Lake	4	11
1918	30 July	Tolland	1	3
1919	21 June	Empress	1	1
1927	8 July	Wetaskiwin	3	Unknown
1950	11 August	Morley	4	6
1960	3 August	Travers Dam	1	3
1961	7 July	Gooseberry Lake	1	1
1972	28 July	Bawlf	1	2
1984	29 June	Richmond Park	1	1

(Hage 1988)

When Can You Expect a Tornado?

Tornadoes have appeared on the prairies as early as February and as late as November. In extreme cases, they have been known to suck up the snow cover. The usual tornado season, however, is April to October, peaking in late June to early July. June is the preferred month for severe tornadoes with wind speeds of more than 180 kilometres per hour. Tornadoes of all types seem to favour July in Manitoba and Saskatchewan, while 90 per cent of tornadoes in Alberta and Saskatchewan that demolish at least one substantial building strike between 1 June and 15 August. Severe tornadoes are less common in April and May. The preferred time of day for tornadoes is 3:00 PM to 7:00 PM. The Regina tornado of 1912 struck on 30 June at 4:50 PM. The Edmonton tornado of 1987 struck on 31 July at 3:01 PM.

Edmonton, 31 July 1987

An unusually hot and humid air mass had settled on the prairies for most of that week, then a cold front swept in. At 2:55 PM on 31 July 1987 a funnel cloud was sighted near Leduc. It re-formed near Beaumont at 3:01 PM and followed a northerly trail across Millwoods in southeastern Edmonton to the North Saskatchewan River valley. It proceeded north along the valley, then turned toward a trailer park in northeast Edmonton. It travelled another five kilometres before dying. It lasted over an hour on the ground, leaving a wake of destruction forty kilometres long and varying in width from less than a hundred metres to over a thousand. Its average speed was thirty-five kilometres per hour.

The Edmonton Tornado—pegged at F3, with damage approaching F4—is the second-worst natural disaster in Canadian history (Hurricane Hazel killed ninety-one people in Ontario in 1954). It obliterated everything in its path—including, tragically, much of Evergreen Trailer Park. Transmission towers were toppled, cars were tossed about like toys, trees were uprooted, and a giant oil storage tank was moved 300 metres. The destructive force of the twister was accompanied by a thunderstorm with hail the size of softballs, winds in excess of 110 kilometres per hour, and forty to fifty millimetres of rain.

Saskatoon, 1 June 1986

A rare multiple tornado event ripped through Saskatoon on 1 June 1986. There were no deaths or serious injuries, but it did cause severe property damage. At 6:45 PM, a line of thunderstorms was building rapidly near North Battleford and moving toward Saskatoon. By 7:00 PM the temperature had risen to 29°C, and the relative humidity was also climbing. A barrage of thunderstorms, lightning, hail, and tornadoes hit Saskatoon at 9:45 PM and lingered until 10:00 PM. According to reports and observations,

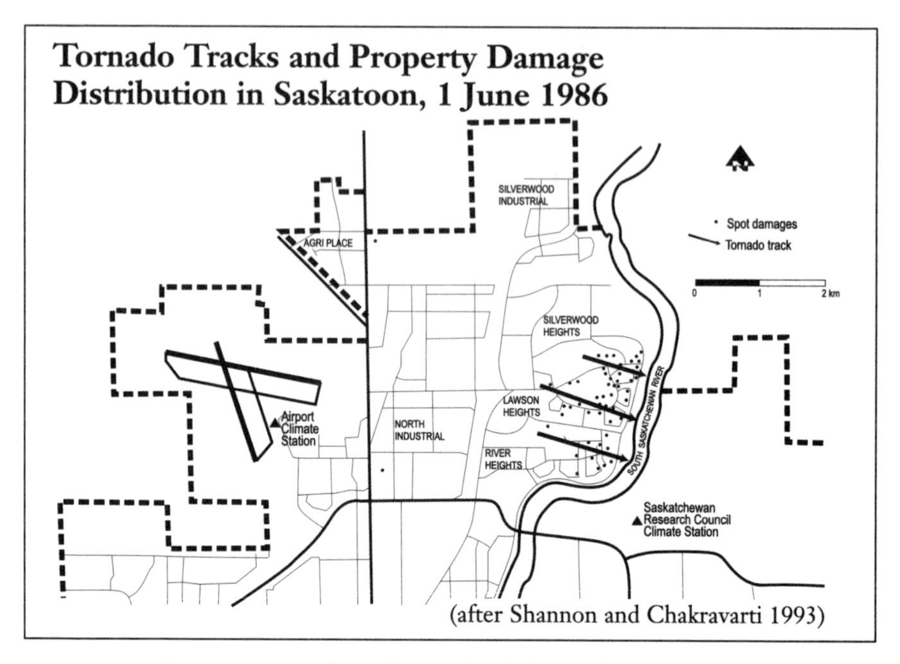

Tornado Tracks and Property Damage Distribution in Saskatoon, 1 June 1986

(after Shannon and Chakravarti 1993)

as many as three tornado funnels touched down that Sunday evening and tracked through the city.

Much of the property damage was minor, for a tornado: shingles ripped off, fences demolished, trees uprooted. Other damage indicated F3 strength as a roof was ripped off a house on O'Brien Crescent, wood was found embedded in a car, and an entire warehouse was obliterated.

A family of tornadoes is unusual enough, but there were other surprising

Too Close to Home

hen I picked up the *Canadian Geographic* magazine for July/August 1996 I did a double take. The cover picture was of a twister spinning through the pastures within a few kilometres of my home south of Saskatoon. A wave of horror swept through me as I saw that devil again. I had watched it from my office in the city, but this photo was taken in a nearby field.

The tornado appeared to be heading straight for Saskatoon, but fortunately it veered eastward some thirteen kilometres from the city limits. No lives were lost, but it was a close call.

Pick and Choose

Tornadoes follow erratic paths, sometimes appearing to wipe out objects selectively and at will. They seem actually to hunt for trailer parks. But how selective are they, really?

My brother Erik was painting a row of granaries one summer when a severe wind storm—he swore it was a tornado—struck. Some of the granaries were picked up and tossed into a nearby field as if they were salt-shakers. Others remained standing. Curiously, it was the unpainted granaries that survived; the tornado had chosen to destroy only those that had recently been painted.

Were there any other witnesses? It sounds to me like the same tornado that pulled fence posts out of the ground and stacked them in a neat pile.

The winds of this tornado drove straw into a post. (Joe Eley)

Tornado Tales

There are many remarkable tales about tornadoes. Destructive though they are, tornadoes have been known to pick up sleeping babies in their carriages and deposit them unhurt back on the ground. In June 1906 near Parkland, Alberta, a house, complete with occupants, was lifted and carried several hundred metres.

A tornado near Sedley, Saskatchewan, turned a cow upside down, anchoring her to the ground with her own horns. She was fine once she was dug out.

Another tornado reportedly lifted a locomotive weighing hundreds of tonnes, then set it down pointing the other way, still on the tracks.

Tornadoes have been reported to pluck live chickens, remove the hair from horses, and drive straw through steel. In the United States a tornado once lifted a railroad coach with 117 passengers and left it in a ditch twenty-five metres away. Another tornado, having demolished a school house, carried the eighty-five students over a hundred metres without any of them being killed.

characteristics about these storms. First, they came out of the northwest, not the south, which is more usual. Second, the tornadoes were triggered by a warm frontal system and a trough, not a cold front, which is more common. Third, the jet stream was present but very weak; the readings of the atmospheric conditions in the upper air over the city showed no warning of possible severe weather conditions. Later that month, on the 26th, another tornado hit the same area, damaging a home that had been struck by one of the earlier twisters. Before 1986, no tornadoes had hit the city for fifteen years.

Regina, 30 June 1912

Sunday was steaming hot, one more day in an extended heat wave. During the afternoon, people began to notice dangerous-looking clouds forming in the southwest. The funnel cloud first touched ground more than eighteen kilometres from the city, turning dark brown as it sucked up soil and debris. Its first victim was the Thomas Beare family farm, where it tossed machinery and animals about as if they were toys. It also battered the family with debris, but they survived.

Five kilometres south of Regina, the tornado sounded like a massive freight train crossing a bridge. It destroyed the Stephenson farm house, and demolished several more farm yards and a church on its way to the city. The first person to die was Andrew Roy from Quebec, who was visiting at the Kerr farm. Two more people were dead before the tornado reached the more densely populated parts of the city at about 4:50 PM.

Bruce Langton and Philip Steele were canoeing on Wascana Lake when the tornado caught them, spinning their canoe into the air. Steele flew out, but Langton somehow stayed with the boat until it landed right-side-up deep in Wascana Park. He was dazed and shaken when found, but he was sitting upright in the canoe with the paddle still clutched in his hands.

A car crashed through the front wall of W. J. Waddell's home on Cornwall Street, bounced off an interior wall, and flew back into the street, barely missing Mrs. Waddell. The family gathered in a corner while their house was flattened. All three family members survived.

There were countless strange events that afternoon: a paper photograph was propelled edgewise into a wall; central sections of bookcases were removed while side portions were left intact; a sliver of wood was found driven deeply into a telephone pole; a piano, stripped of its inner parts, was found in the street; a baby was found huddled in an oven far from any kitchen. Papers and books were strewn in all directions and lost in the storm, including the examination papers for all Saskatchewan grade schools, which had been piled on a table in the Department of Education.

One full block of residences between Lorne and Smith Streets, bounded by 13th and 14th Avenues, was directly in the path of the tornado and destroyed. Amazingly, there was only one fatality on the block; a baby, Donald Miller Loggie, was killed instantly when his home was smashed. In

Tornado Do's and Don'ts

If you're in a vehicle, don't try to outrun the twister. Above all, don't try to chase it. Get out of the car and move well away from it so it won't roll over you or be dropped on you. Lie down in a ditch, cover your head, and stay there. Avoid drainage ditches or dry creek beds, as they often flood during a storm.

all, 500 buildings were damaged or demolished. The path of destruction was 300 to 400 metres wide. Moving at fifteen to eighty kilometres per hour, the twister continued for at least another twelve kilometres after it left Regina.

The aftermath was confused and chaotic. Financing the rebuilding was a problem, and it was not until forty-seven years later that the last of the Province's loan was repaid. However, there was hardly any looting or vandalism, and practically all of the wreckage was cleaned up.

References

Adam, B. A. 1992. Frost, snow take toll on some Sask. crops. Saskatoon *Star Phoenix*, 24 August.

Ahrens, C.D. 1993. *Essentials of meteorology: An invitation to the atmosphere*. St. Paul, MN: West Publishing Co.

Anderson, F. W. 1968. *Regina's terrible tornado*. Calgary: Frontier Publishing.

Booth D. 1996. *The dust bowl*. Toronto. Kids Can Press Ltd.

Bragg, R. 1992. If only politicians were like weather. Calgary *Herald*, 26 August.

Calgary *Herald*. 1982. Freak snowstorm strikes, 31 May.

______. 1992. It's the 46-year blues as snow hits, 22 August.

Diefenbaker, John G. 1975. *One Canada: Memoirs of the Right Honourable John G. Diefenbaker, Vol. I: The crusading years 1895–1956*. Toronto: Macmillan of Canada.

Department of Transport (DOT), Meteorological Division. n.d. *Monthly record of meteorological observations in Canada and Newfoundland, January 1947*. Toronto: Department of Transport.

Environment Canada. 1990. *Severe weather safety*. Ottawa: Environment Canada.

———. 1991. Tornado outbreak in the prairies. *Climatic Perspectives* 13(22):1.

______. 1992. Surprise snowfall blankets Alberta. *Climatic Perspectives* 14(34):17–23.

______. 1992. *Monthly meteorological summary, August 1992 at Calgary, Alberta (Airport)*. Calgary: Environment Canada.

______. 1993. *Canadian climate normals on diskette*. Version 2.0E. Downsview, ON: Environment Canada.

Environment Canada, Atmospheric Environment Service. 1978. *Monthly record, meteorological observations in western Canada, February 1978*. Downsview, ON: Environment Canada, Atmospheric Environment Service.

Etkin, D., and A. Maarouf. 1995. An overview of atmospheric natural hazards in Canada. In *Proceedings of a tri-lateral workshop on natural hazards, Sam Jakes Inn, Merrickville, Canada, Feb. 11–14, 1995*, edited by D. Etkin, 1–63 to 1–92. Downsview, ON: Environment Canada.

Garard, D. 1992. Snow buries farmers' hopes. *Calgary Herald*, 23 August.

Grazulis, T. P. 1991. Significant tornados, 1880–1989. Vol. 1. Discussion and analysis, Environmental Films, St. Johnsbury, VT. In *Beyond the year 2000, more tornados in western Canada? Implications from the historical record*, edited by D. Etkin. In *Natural Hazards* 12:19–27.

Hage, K. D. 1988. A comparative study of tornados and other destructive windstorms in Alberta and Saskatchewan. In *The impact of climate variability and change on the Canadian prairies: Symposium/workshop proceedings, 9–11 September 1987*, edited by B.L. Magill and F. Geddes, 351–77. Edmonton: Alberta Department of the Environment.

Huschke, R. E., ed. 1959. *Glossary of meteorology*. Boston: American Meteorological Society.

Klein, G. 1984. Two die in accident. Saskatoon *Star Phoenix*, 1 June.

Knisley, J. 1982. Freak May snow leaves southwest battered and buried. Regina *Leader Post*, 31 May, p. A1.

Kogan, F. 1995. Advances in using NOAA polar-orbiting satellites for global drought watch. *Drought Network News* 7(3):15–20.

Lanken, D. 1996. Funnel fury. *Canadian Geographic* (July/August):24–31.

Maple Creek News. 1982. Remembering the storm, 1 June.

______. 1982. What a snow storm!! 1 June.

Nanton and District Historical Society. 1975. Mosquito Creek roundup. Nanton, Alberta. In Hage, K. D. 1988. *A comparative study of tornados and other destructive windstorms in Alberta and Saskatchewan*. In *The impact of climate variability and change on the Canadian prairies: Symposium/workshop proceedings*, 9–11 September 1987, edited by B.L. Magill and F. Geddes, 351–77. Edmonton: Alberta Department of the Environment.

Paul, A. 1994. Tornadoes and hail. In *Proceedings of a workshop on improving responses to atmospheric extremes: The role of insurance and compensation. Toronto, 3 and 4 October*, edited by J. McCulloch and D. Etkin.

______. 1995. The *Saskatchewan tornado project*. Regina: Department of Geography, University of Regina.

Paul, A., and D. E. Blair. 1993. The thunderstorms of 8 July 1989 in the northern Great Plains. *Climatological Bulletin* 25:47–59.

Phillips, D. 1990. *The climates of Canada*. Ottawa: Canadian Government Publishing Centre.

______. 1993. *The day Niagara Falls ran dry! Canadian weather facts and trivia*. Toronto: Key Porter Books.

______. 1995. *Weather flashes: Canada's weather trivia mini-magazine*. Saskatoon: Fifth House Publishers.

Raddatz, R. L., and J. M. Hanesiak. 1991. Climatology of tornado days 1960–1989 for Manitoba and Saskatchewan. *Climatological Bulletin* 25:47–59.

Regina *Leader Post*. 1947. 31 January and 3 February.

______. 1978. 7 and 8 February.

______. 1982. Snow in Swift Current, 28 May.

______. 1982. This is spring? (photo caption), 29 May.

______. 1992. Central Sask. suffers through August snowfall. Special from the *Star Phoenix*, 22 August.

______. 1992. Early winter costly—Alberta crops may be ruined, 24 August.

Rocky Mountain House Reunion Historical Society. 1977. *The days before yesterday: History of Rocky Mountain House District*. Rocky Mountain House, AB.

Saskatchewan Agriculture. 1992. *Crop and weather report*. Report #21 for 24 August 1992. Regina: Economics Branch, Saskatchewan Agriculture.

Saskatoon *Star Phoenix*. 1982. What happened to spring? 29 May.
______. 1982. Farmers welcome additional moisture, 31 May.
______. 1982. Spring storm creates havoc in province, 31 May.
______. 1992. Snow in August illustrates weird weather on prairies, 22 August.
______. 1992. Snow dumped on Alberta "abnormal," 24 August.
______. 1992. Cold hurts Sask. crops, 25 August.
Saskatoon *Sun*, 1996. 22 December.
Shannon, R. E., and A. K. Chakravarti. 1993. The June 1986 tornado of Saskatoon: A prairie case study. *Prairie Forum* 18(2):269–78.
Skarpathiotakis, M. 1991. Edmonton tornado leaves a trail of death and destruction. *Climatic Perspectives* 9(31).
Swift Current *Sun*. 1982. Snow storm staggers southwest—blizzard plays havoc in city, 1 June.
______. 1982. Blizzard leaves district in chaos, 2 June.
______. 1982. Ag-rep says area cattle suffering from malnutrition after storm, 3 June.
______. 1982. Grain companies say grasshopper problem eliminated by blizzard, 3 June.
______. 1982. Piapot man dies as result of storm, 3 June.
______. 1982. Worst May weather in decades research station officials say, 3 June.
______. 1982. Blizzard in Gull Lake worst in town's history, 3 June.
______. 1992. Rain, frost delay harvest in southwest, 26 August.
Wallace, A. 1987. The Edmonton tornado, July 31, 1987. *Climatic Perspectives* 9:7b–8b.
Wheaton, E. 1992. Prairie dust storms—a neglected hazard. *Natural Hazards* 5:53–63.
Wheaton, E., A. Chakravarti, and V. Wittrock. 1993. *The sensitivity of dust storm hazards to climatic conditions on the Canadian prairies*. American Association of Geographer's Annual Meeting, Atlanta, GA.

Chapter Two

Weather in Time: Seasons

The drama of this landscape is in the sky, pouring with light and always moving . . . It is a long way from characterless; "overpowering" would be a better word. For over the segmented circle of Earth is domed the biggest sky anywhere, which on days like this sheds down on range and wheat and summer fallow a light to set a painter wild, a light pure, glareless, and transparent.

Wallace Stegner, *Wolf Willow*

The prairie climate is characterized by change. Enormous swings of temperature and precipitation bring us winters of paralysing cold and summers of blistering heat. Spring and fall, too, are not always times of mild transition. They can bring brutal changes as well. Sometimes it seems as if winter and summer have occurred on the same day, perhaps within hours of one another.

The prairie is a battlefield of opposing air masses. Monstrous wars are fought here, for this is where bone-dry, bitterly cold air masses descend from the Arctic to do battle with moist, tropical air moving up from the south. When Arctic air dominates, the prairies experience the lowest temperatures on the continent. When continental tropical air masses roll in, prairie temperatures are the hottest and the air is the driest. The clash of air masses, encouraged by the north-south alignment of the mountain ranges in North America, can result in violent hail storms, blizzards, intense rains, and tornadoes. They also account for the year-to-year, day-to-day, sometimes hour-to-hour changes in the weather.

As we were taught in elementary school, seasons occur because the earth revolves around the sun. During the summer, the northern hemisphere is tipped toward the sun, and the most solar energy is received. During the winter, the northern hemisphere is pointed away from the sun, and receives much less energy. Prairie seasons, however, often arrive and depart at quite different times from year to year, so defining the seasons can be a challenge. Climatologists conventionally classify December, January, and February as winter; March, April, and May as spring; June, July, and August as summer; and September, October, and November as fall. Other definitions are based on astronomy or temperature. Winter, for example, is sometimes defined as beginning when average daily temperatures drop below 0°C over a given area (a less than useful definition in tropical climates). Astronomically, summer can be classified as beginning at the summer solstice on 21 or 22 June, which has the longest period of daylight in the year. Winter, consequently, would begin six months later at the winter solstice, 21 or 22 December, the longest night of the year—known as "Mother Night" among the ancient Celts, for it was thought that this night gave birth to all the other nights of the year.

A major factor controlling prairie climate and the flow of the seasons is location: the great plains are located at the heart of an enormous continent, far from any oceans. Nearer the ocean, there is much less seasonal variation in temperature. Water heats and cools more slowly than land. These areas are said to have a maritime climate. The prairies have a continental climate.

Summer-to-winter temperature differences are greater on the prairies than in any other part of Canada. Compare the seasonal swings of temperature at Winnipeg and Vancouver. Vancouver, with its maritime climate, has warmer winters and cooler summers, whereas Winnipeg, with its

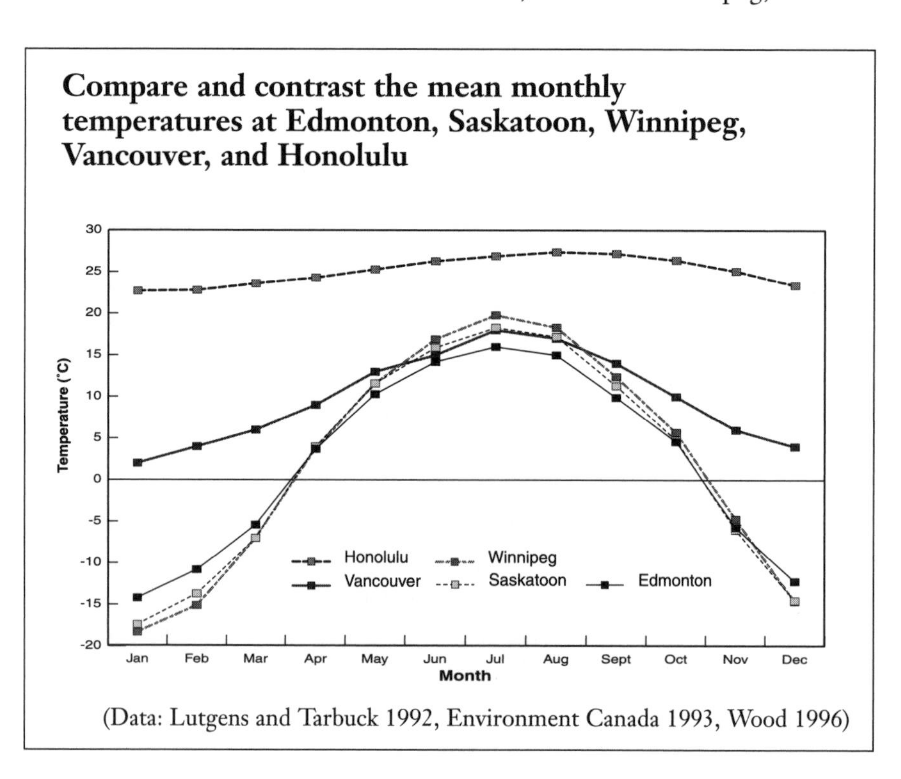

Compare and contrast the mean monthly temperatures at Edmonton, Saskatoon, Winnipeg, Vancouver, and Honolulu

(Data: Lutgens and Tarbuck 1992, Environment Canada 1993, Wood 1996)

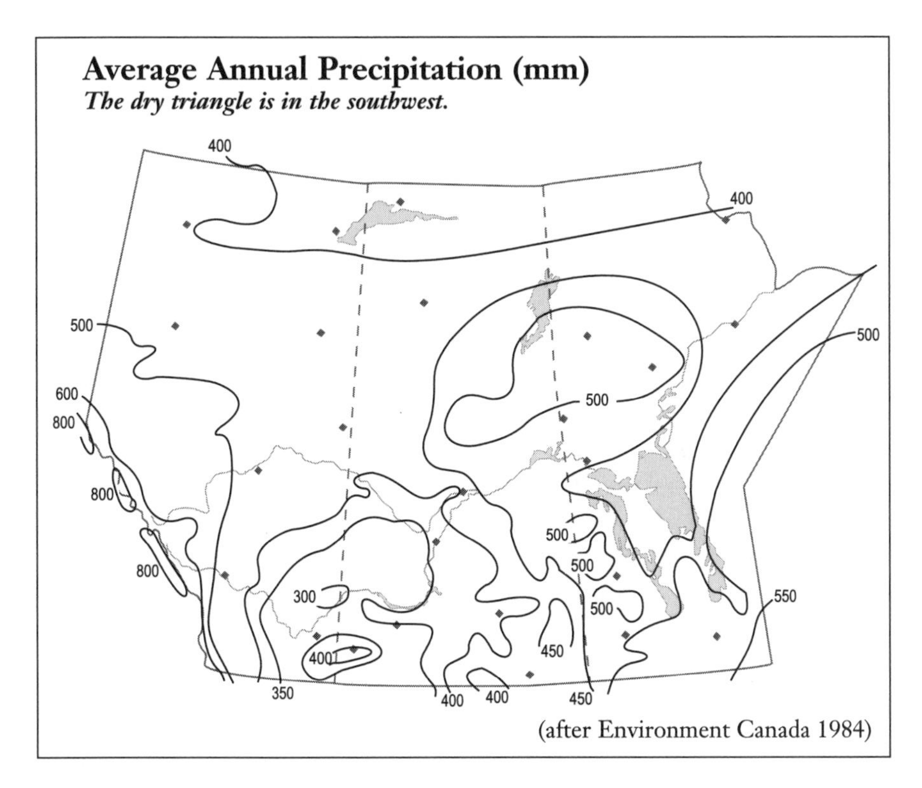

continental climate, has a mean January temperature much lower than Vancouver's, but the summers are warmer.

For a more dramatic contrast, let's consider the mean temperatures of Honolulu, Hawaii. The mean January temperature in Honolulu is about 23°C; the mean July temperature is about 27°C. Tropical maritime locations tend to have greater day-to-night than season-to-season differences in temperature.

A recent and vivid example of the weather extremes the prairie dweller contended with was the year 1996. Spring brought wet, cold weather, with accompanying floods to Manitoba and eastern Saskatchewan. August, in contrast, brought drought to some parts of the prairies, especially northwestern agricultural Saskatchewan and central agricultural Alberta. Then in September, rains delayed the harvest, and before the harvest was over, snow started to fall.

In general, though, the prairies are drier than any other region of Canada except the Arctic. The driest area of the prairies is the Palliser Triangle, which takes in portions of southwestern Saskatchewan and southeastern Alberta. This dry belt averages 350 millimetres of precipitation annually.

The heart of the belt averages less than 300 millimetres, and desert vegetation and animals are common. The area is, in fact, a northward extension of the Great American Desert, which takes in a number of the southwestern United States and averages less than 250 millimetres of precipitation annually. The world does have drier places, though, and the prairies are listed only among the semi-arid regions of the world.

Much wetter areas, with average annual precipitation over 500 millimetres, flank the dry belt to the west and to the east. To the north, the western boreal forest is drier than its eastern counterpart, with precipitation averaging less than 400 millimetres toward the North West Territories to more than 500 millimetres north of Lake Winnipeg.

Despite these extremes, the Canadian prairie is generally a land of clear skies and bright sunshine. The prairies receive by far the most sunshine of any region in Canada, with southern Saskatchewan receiving the greatest number of sunshine hours per year on average (over 6.5 hours a day). Here the annual number of bright sunshine hours often exceeds 2,400. An area

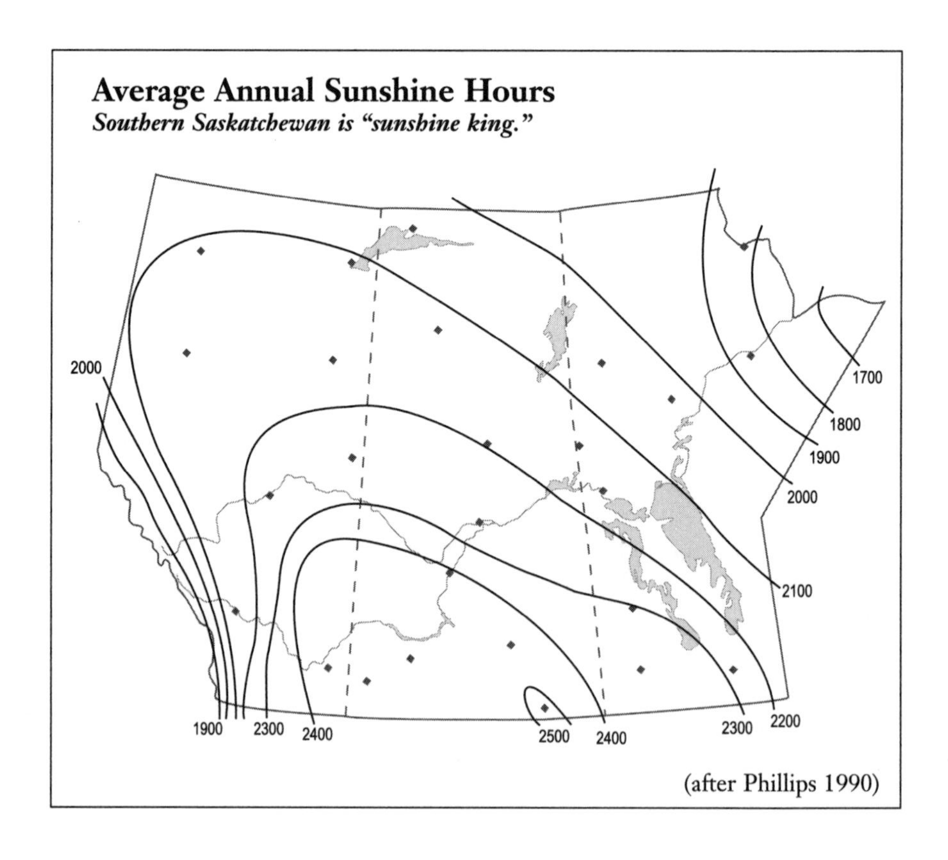

along the coast north of Prince Rupert, BC, in contrast, receives less than 1,000 hours of sunshine per year on average.

The most sunshine records are held by Saskatchewan, followed by Alberta. Estevan, Saskatchewan holds the record for the highest average annual hours of sunshine at 2,537. Saskatoon is the major city with the greatest average annual hours of sunshine, at 2,450. Regina is the sunniest provincial capital, at 2,331 hours annually. Manyberries, Alberta scores with the sunniest year on record: 2,785 hours.

During the summer, the sun has control of the weather. Daily temperature trends show regular upward tendencies during the day and downward tendencies during the night. In summer, the highest temperatures usually occur in the late afternoon, with the lowest temperatures late at night, or early morning, near sunrise.

Winter, on the other hand, brings surprises. The temperature can drop during the day and increase during the night. This is because in winter the weather is largely controlled by the movement of air masses across the continent. A low pressure system with a warm triangular area can bring rising night-time temperatures. If a high pressure system follows on its heels, we can expect extremely cold conditions to follow.

Winter

Prairie dwellers do not have to dream about white Christmases. We usually have them. And they do make the holidays brighter, with sunlight or moonlight reflecting off the snow and ample snow cover for recreation. But, as we all know, prairie winters are a challenge, if not downright dangerous. Our lowest temperatures are expected in mid to late January, but snowstorms and low temperatures can strike anywhere from the end of August (as in 1992) to the end of May, even later in the north (as in 1982).

January occupies the low spot on the temperature chart. It is the month with the lowest average winter temperatures on the prairies, ranging from highs around -10°C in balmy southwestern Alberta to lows around -27°C across the northeast.

The January temperature isotherm patterns slant from northwest to southeast. This means that places a similar distance north of the United States border are cooler on average as you move eastward, and vice versa. Prince Albert and Edmonton, for example, are at 53°13' N and 53°18' N, but Prince Albert's winter temperatures average almost 6°C lower than Edmonton's.

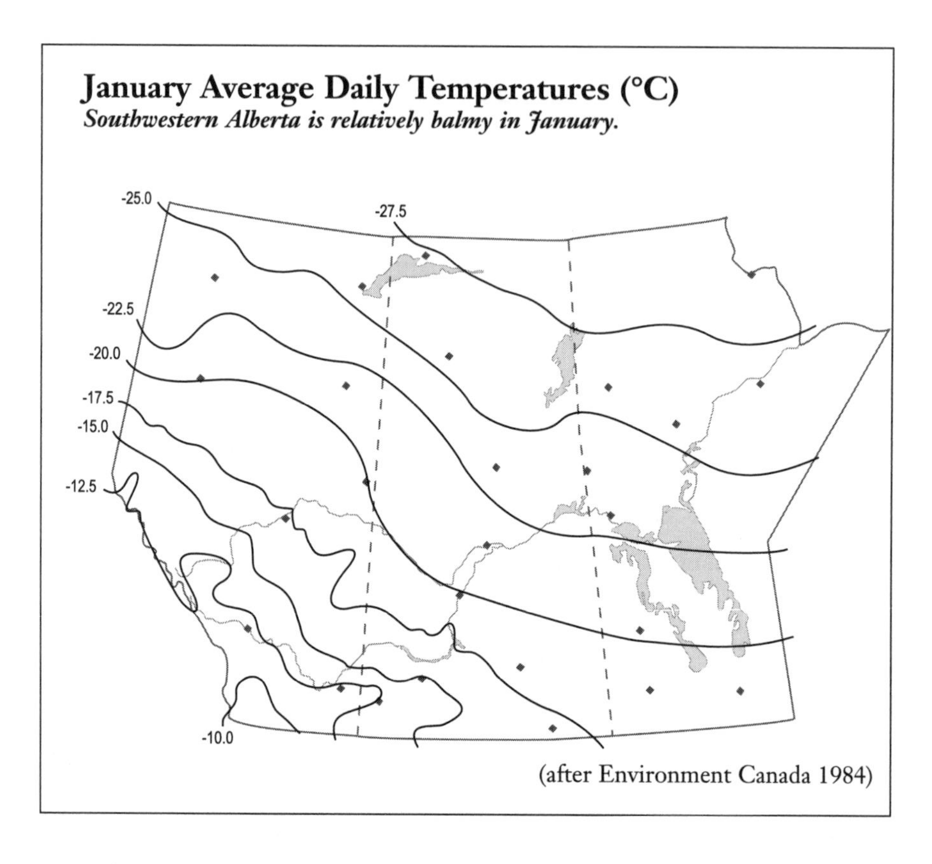

January Average Daily Temperatures (°C)
Southwestern Alberta is relatively balmy in January.

(after Environment Canada 1984)

Isotherms show the patterns of temperatures across a given area. The map of January average temperatures shows lines labelled -25°C and -20°C. Moving northeast from the -20°C line past the -25°C line, lower and lower temperatures are expected. The region between these two lines will usually have average January temperatures between these values. Daily average highs in January range from around -25°C in the northeast to around -5°C in the southwest. Daily average low temperatures range from around -30°C in the extreme northeast to around -15°C in the extreme southwest.

The isotherms are also tightly spaced in winter; temperatures decrease northeastward much more rapidly than at any other time of the year. This means that the average temperature difference between two places along a northeast to southwest line is much greater in winter than it is during the summer. So don't expect northern locations to have a substantially cooler summer than

A Winnipeg blizzard, 5 April 1997. (Joe Bryksa, Winnipeg *Free Press*)

places a few hundred kilometres to the south. But do expect larger winter differences.

Surprisingly, the coldest places contradict the provincial averages in terms of ratings. Fort Vermilion, in balmy Alberta, holds the record low temperature on the prairies of -61.1°C, set on 11 January 1911. Prince Albert, Saskatchewan, is next lowest at -56.7°C, recorded on 1 February 1893. At -52.8°C, Manitoba holds the least severe of the record low temperatures, set at Norway House on 9 January 1899. Note that these records occurred early in the station's history, a trend that holds for most of Canada. Temperatures were really low "in the old days."

The Coldest Cities

Winnipeg's January mean temperature of -18.3°C ranks barely below Saskatoon's -17.5°C and Regina's -16.5°C. Edmonton ranks next at -14.2°C, while Calgary is a much warmer -9.6°C. So Winnipeg wins the award for the coldest city on the prairies. But which city *feels* the coldest? To determine that, wind-chill must be taken into account, in which case

Significant Events in the History of Cold Weather in Canada

- The lowest temperature ever recorded in Canada was -63°C at Snag, Yukon, on 3 February 1947.
- Starting on 24 January 1966, Winnipeg recorded the longest period of skin-freezing wind-chill lasting 170 consecutive hours.
- The temperature at Edmonton remained below -18°C from 7 January until 2 February 1969, a long cold spell of twenty-six days.
- The coldest wind-chill recorded in Canada occurred at Pelly Bay, NWT, on 13 January 1975. The wind-chill was the equivalent of -92°C, or 3357 w/m^2 (watts per square metre).
- The longest period with wind-chill recorded at Saskatoon began on 28 December 1978 and lasted 215 hours.
- January 1982 was one of the coldest months recorded in Canada, with temperatures below -40°C in most provinces.
- A freeze in British Columbia on 30 January 1989 resulted in burst water pipes across the province, which in turn resulted in losses of $2.5 million.
- Cold water, causing hypothermia, was responsible for seventy-seven of the 639 drowning fatalities in Canada in 1991.

(after Etkin and Maarouf 1995)

Regina takes the prize. The average January wind speed in the Saskatchewan capital is twenty-one kilometres per hour—appreciably higher than Winnipeg's eighteen.

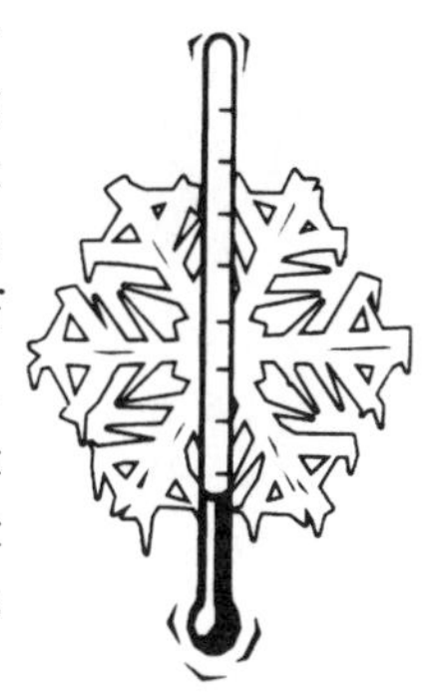

The record low—as opposed to mean or average—temperatures for these cities tell another tale. Winnipeg is not the coldest, and Calgary is not the warmest—at least, not the only warmest. Regina and Saskatoon are both "winners," each having recorded a record low of -50.0°C, and Edmonton runs a close second at -48.3°C. Winnipeg and (surprise!) Calgary are tied for third at about -45.0°C each. Winnipeg may have the lowest average January temperature, but Regina's extremes are greater.

The coldest region on the prairies, northeastern Manitoba, experiences the most days with temperatures below -20°C—on average, more than 120 days per year. The warmest region, southwestern Alberta, has the fewest

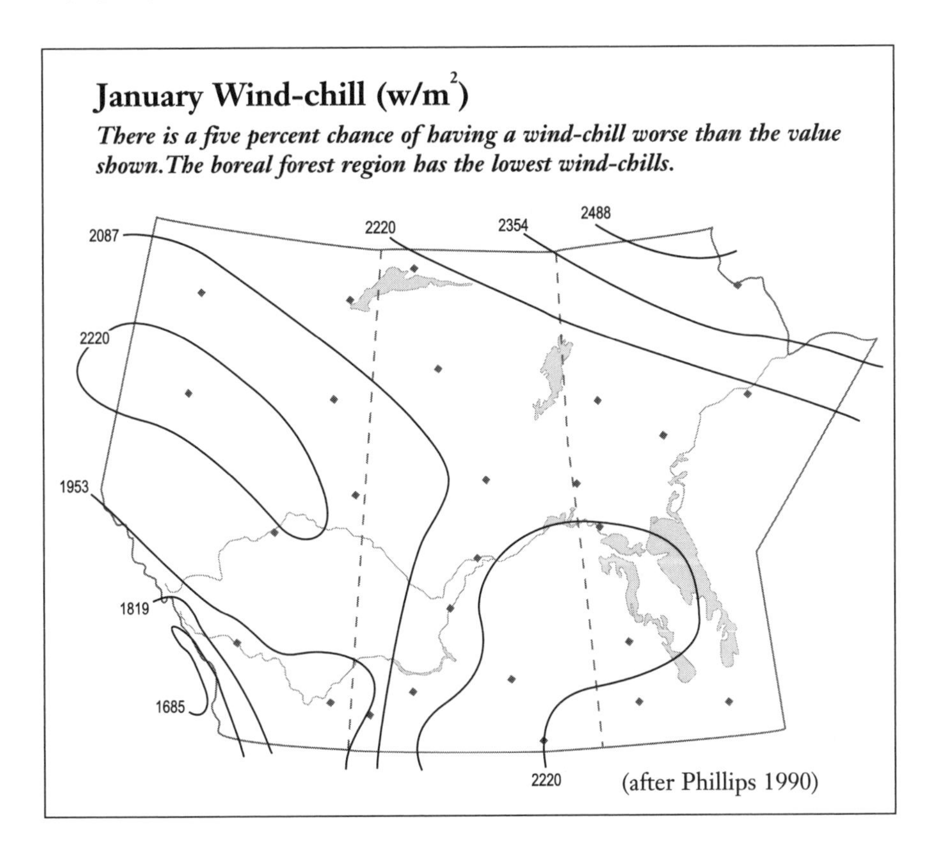

What's Missing?

The morning of 26 November 1995 was cool and clear. I went out to feed the animals. As I was working, I got the odd feeling that something was missing. I counted the heifers, but they were all there. I looked around with a suspicious eye. Then I used my ears, and found it! The roar of the wind was gone. The morning was quiet. I could hear the birds singing, the cattle chewing, and the snow crunching beneath my boots. It had been a long time since I had heard the sound of silence out here on the prairies.

cold days—less than thirty days below -20°C. The difference between the cold northeast and the warm southwest is ninety to a hundred days on average.

The Bonspiel Thaw

Just when prairie people are fed up with the cold and desperately in need of a change, something happens. No, we don't go south. The south comes to us, as a bonspiel thaw, a period of unusually warm weather that occurs

Average January and July and Extreme Temperatures (°C) for Major Prairie Cities

City	January Average (1961–90)	Extreme Low (to 1991)	July Average (1961–90)	Extreme High (to 1991)
Edmonton	-14.2	-48.3 Jan. 26, 1972	16.0	35.0 July 14, 1961
Calgary	-9.6	-45.0 Feb. 4, 1893	16.4	36.1 July 25, 1933
Saskatoon	-17.5	-50.0 Feb. 1, 1893	18.6	40.6 June 5, 1988
Regina	-16.5	-50.0 Jan. 1, 1885	19.1	43.3 July 5, 1937
Winnipeg	-18.3	-45.0 Feb.18, 1966	19.8	40.6 Aug. 7, 1949

(Data: Environment Canada 1993)

Toboggan flights. (John Perret)

at similar times each year, usually mid to late January. This is when a westerly moving mild Pacific air mass brings spring-like conditions to the frozen prairie. The thaw can last from a few hours to a week, but it comes most winters in the western prairies. In fact, the likelihood of having one or more January days with above-zero temperatures is greater than 90 percent in the Edmonton area, though it decreases as you move east—to around 80 percent in the Regina area and about 58 percent around Winnipeg.

This singularity in the weather is called the "bonspiel thaw" because it often occurs when curling bonspiels are in full swing. It comes as a great relief after weeks of -30°C or -40°C weather. But the thaw can cause problems. Diehard curlers actually used to complain about it because it ruined the ice (this was in the days before artificial ice). An unseasonal thaw can cause plants to emerge from their dormant period prematurely, and subsequently fall victim to winter kill. An unseasonable reduction in snow cover can also result in the ground freezing deeply in ensuing cold spells, having lost the insulation value of deep snow cover. It just goes to show that you can't please everybody, especially when it comes to the weather.

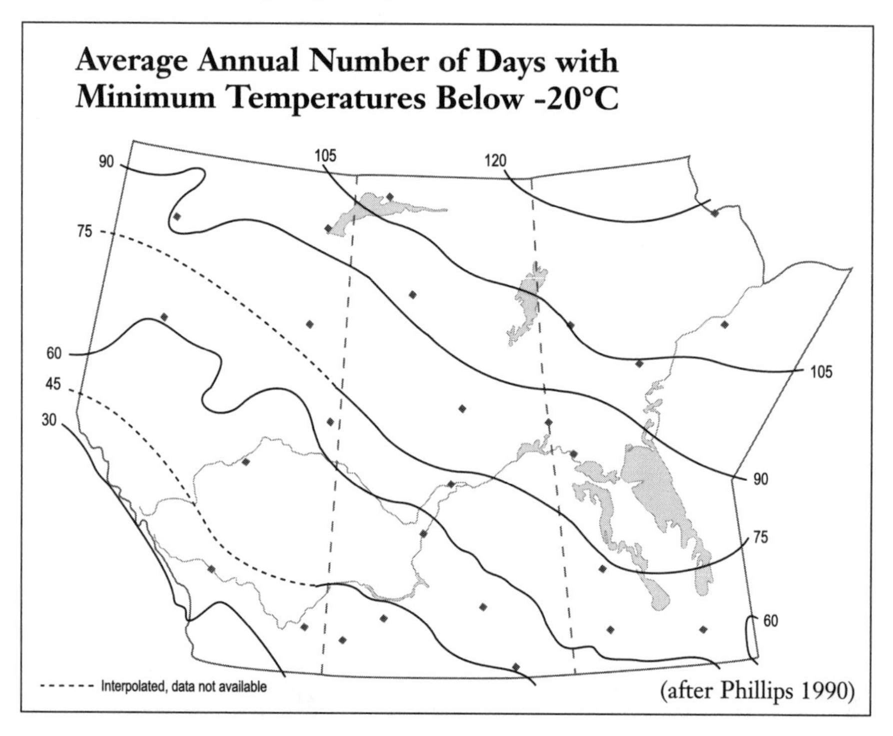

(after Phillips 1990)

Wind-chill

Wind-chill is not a harsh enough word for the conditions we have to survive. Wind-"chill" should be wind-"freeze" or wind-"paralyze" because it is never just a "chill." After all, you can have summer wind-chill values. And even at -30°C, if the sun is shining and you're in a good mood, the weather may not seem too bad if you're dressed for it. But when the wind rises, even 0°C seems intolerable. The stronger the wind, the colder it feels, even if the temperature hasn't changed. This is because the wind removes your body heat more quickly and evaporates moisture on your skin.

Not Cold Enough?

It was only -38°C at seven o'clock on the morning of 18 January 1996. When I woke up my teenage sons, the first thing they asked was, "What's the temperature?" No, they were not intent on a new science project, though they had studied the school closure information and the weather forecast the night before. If the temperature at 7:00 AM is -40°C the school bus is cancelled. It also doesn't run when the wind-chill is greater than 2200 w/m^2, visibilities are poor, or if the roads are slippery. When they learned that it was *only* -38°C, they groaned in disappointment.

The next morning, the boys got their wish. The temperature was -40°C at 7:00 AM, and the bus operator phoned to say that he was not driving. I wish I could have spent the morning in bed to escape that bitterly cold weather.

Wind-chill is a measure of the cooling effect of the combination of temperature and wind, expressed as the loss of heat in watts per square metre (w/m^2). It is an indication of how fast an object will cool to the temperature of the surrounding air, and how difficult it is to maintain the object at a temperature above that of the surrounding air. The rate of cooling can only be approximate, though, because it changes with the shape of the body and other factors such as humidity and sunshine.

Of course, wind-chill increases with greater wind speeds. A temperature of -35°C with a light wind of ten kilometres per hour (1,730 w/m^2 wind-chill) seems warmer than -10°C with a strong sixty kilometre-per-hour wind (1,782 w/m^2 wind-chill).

The worst wind-chills on the prairies occur in south-central Saskatchewan and along a corridor in northern Alberta, where there is a five percent chance that January wind-chills will be worse than 2220 w/m^2. In terms of extreme wind-chill days, Churchill, Manitoba scores the highest at 2,938 w/m^2. Wind-chills are even worse over the Barrens northwest of Hudson Bay; these rate as the highest in Canada. The most severe wind-chill recorded in Canada—3357 w/m^2—occurred at Pelly Bay, North West Territories, on 13 January 1975.

Winds in January, the coldest month of winter, are often deadly. Mean wind speeds peak at more than twenty-five kilometres per hour in southwestern Saskatchewan. The Churchill area in northeastern Manitoba comes a close second, with average January wind speeds exceeding twenty-four kilometres per hour. Northern Alberta appears to have the weakest winter winds on the prairies.

Wind-chill User Guide

Wind-chill (w/m^2)	Effect
1400–1600	Normal winter clothing is adequate, but conditions for most outdoor activities cease to be pleasant.
1700–1900	Warm winter clothing is recommended, including facial protection for most outdoor activities. Flesh will freeze with prolonged exposure.
2000–2200	Warm winter clothing is essential. Even with facial protection, outdoor activities such as skiing and walking are not recommended. Travel without adequate precautions is not recommended. Exposed skin will freeze in one to three minutes.
Above 2300	Facial protection is essential. Travel and all outdoor activities can be extremely dangerous. Avoid going outdoors. Skin will freeze in less than one minute.

The Beaufort Wind Scale . . . Somewhat Revised

Force	Beaufort Description	Revised Description
0	Calm	Children want Dad to fly new kite
1	Light airs	Reading newspaper outdoors becomes a problem
2	Light breeze	Reading newspaper outdoors becomes impossible
3	Gentle breeze	Twigs by front window begin to tap on glass
4	Moderate breeze	Side gate bangs in night if someone forget to bolt it
5	Fresh breeze	Old gentlemen's hats blow away
6	Strong breeze	Clotheslines come down, dragging newly washed sheets on grass
7	High wind	Side gate bangs in night, even when bolted
8	Gale	Car steering seems to have gone wrong
9	Strong gale	Old ladies' hats blow away
10	Whole gale	Clotheslines with newly washed sheets take off
11	Storm	Old ladies and old gentlemen blow away
12	Hurricane	Children want Dad to fly new kite

(after Ball 1989)

Miraculous Recovery from Wind-chill Freezing

In late February 1994, a retreating blizzard had sucked Arctic air into the southern prairies. In Rouleau, Saskatchewan, two-year-old Karlee Kosolofski wandered out her front door about 2:00 AM and was accidentally locked outside. The wind-chill froze her flesh within thirty seconds. Her core body temperature fell to 14°C—24°C below normal. Her mother found her six hours later. The little girl was rushed to hospital in Regina, where she was brought back to life by warming her blood with a heart and lung machine. She had survived the lowest body temperature ever recorded. Her injuries were serious, even so: she lost her left leg below the knee to frostbite, and also suffered minor frostbite to her nose, ears, and elbow. Her survival and recovery made world news.

Ironically, it may have been the extreme cold and wind-chill that saved Karlee's life. As the body freezes, water in the tissues forms ice crystals that damage blood vessels. Rapid freezing forms smaller crystals, and therefore does less harm.

This case is a reminder that cold kills more often than any other weather phenomenon, including heat waves, lightning, tornadoes, and floods. About eighteen people in Canada die of cold every year, whereas seventeen people a year die from all other weather events. Frostbite also claims many injuries.

Snow Is Just Dry Rain

Winter is the driest season on the prairies. Only twenty to thirty per cent of the total yearly precipitation falls as snow, but it's more noticeable than rain because it accumulates through much of the season. The least snow falls in south-central Saskatchewan, which averages less than 100 centimetres per year. Amounts increase in all directions from this dry core, especially westward to the foothills, which receive more than 600 centimetres in some locations.

In Canada, British Columbia holds all the records for annual, monthly, and daily snowfall. The greatest snowfall in one season was 2,446.5 centimetres at Revelstoke/Mount Copeland, in 1971–72. The least snowfall recorded is at Eureka, North West Territories, which averages sixty-four millimetres a year. The greatest prairie snowfall, in contrast, was at the Columbia Icefield in Alberta, where 169.2 centimetres fell over five

The Teenage Guide to Winter Dressing

"It's thirty below out here today! Wear something warmer than that."
"This bomber jacket's perfectly warm."
"All right, but take a hat."
"Hats are for total geeks, Mom. I can't wear that to school."
"Well, wear your nice new boots at least."
"No way, 'cause boots aren't cool."
So out she went — no scarf or mitts!
And started the walk to school,
Shivering, frozen, chilled to the bone
But totally, totally cool.

Marion Young 1996

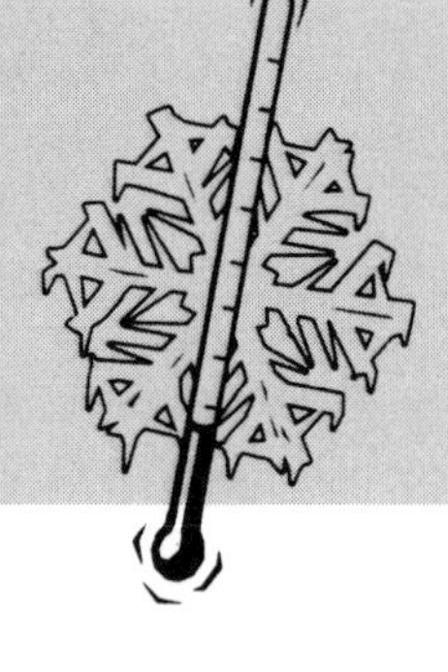

Average Annual Snowfall (cm)

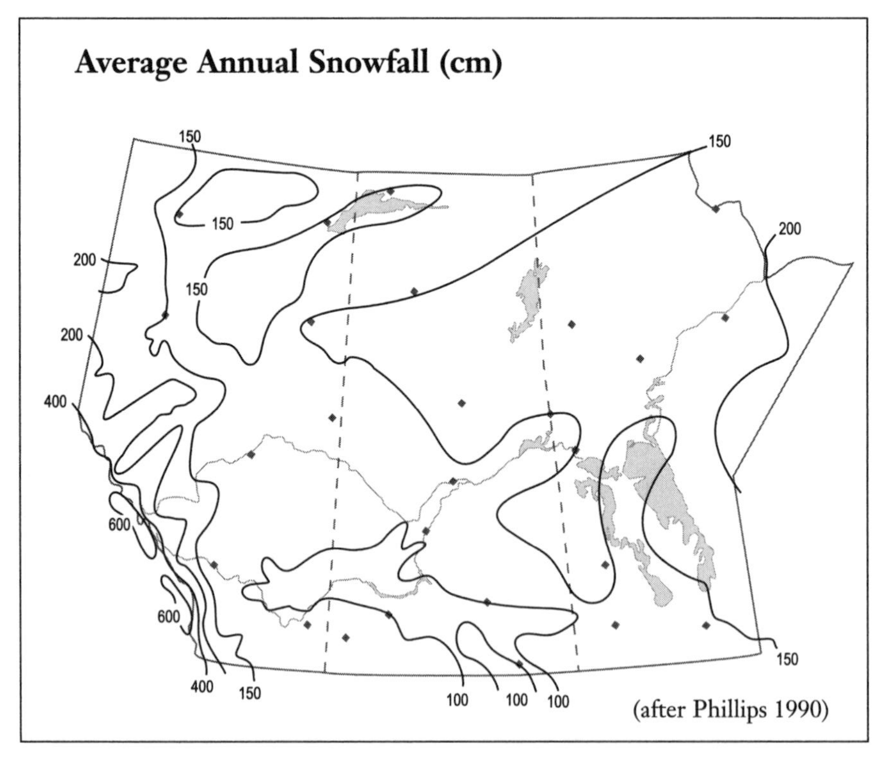

(after Phillips 1990)

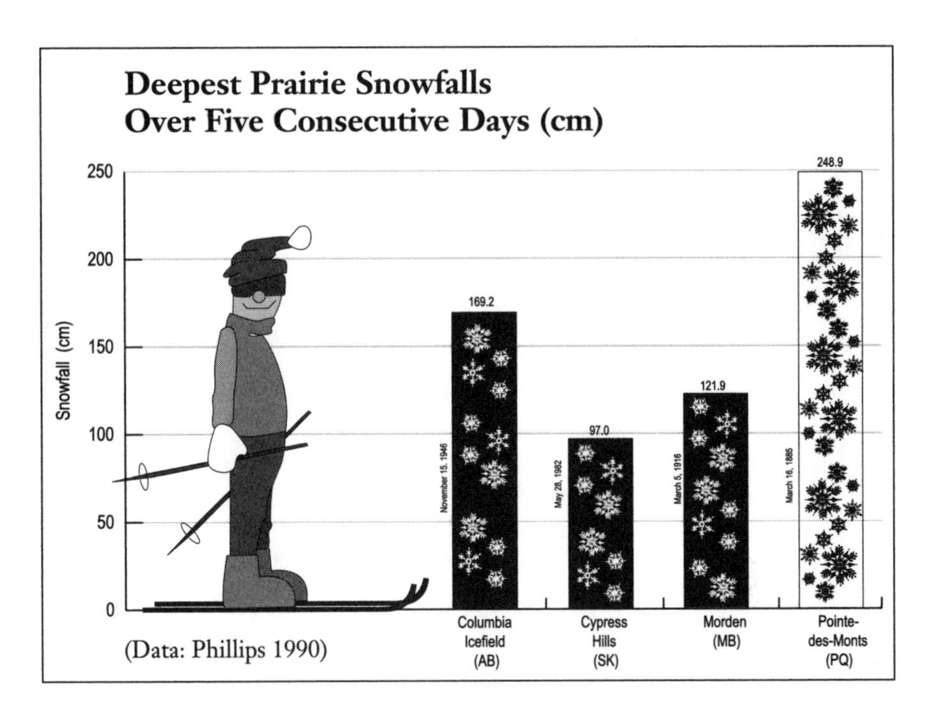

consecutive days in November 1946. Even this was no match for the Quebec snowfall of 248.9 centimetres in 1885.

The depth of the snow cover measures less than the actual amount of snowfall, as snow settles and compresses, sublimates, and blows away. Snow cover on the prairies parallels snowfall patterns: it is lowest in the south central prairies, where the average maximum is less than thirty centimetres, and deepest in the foothills, at over 300 centimetres. The prairies have very little snowfall and snow cover compared to other places in Canada, but the ubiquitous wind can drive it over long distances and pile it up in mountainous drifts that have been known to bury trains and entire buildings.

Saskatchewan holds the prairie record for the most snow on the ground—a surprising statistic, since the province also has the lowest snowfall. Again, the answer is in the wind, as strong winds collect snow from upwind areas. Hudson Bay, Saskatchewan, once recorded 224 centimetres of snow on the ground. Alberta's record is 179 centimetres at Parker Ridge; Manitoba's is 175 centimetres at Glenlea. The Canadian record is 450 centimetres at Whistler Roundhouse, British Columbia. Outside these extremes, the deepest snow cover is usually found in southwestern Alberta, an area that averages deepest snow covers exceeding 100 centimetres.

Politics—Blame It on the Snow

The winter of 1973–74 was severe, with considerable snow on the prairies. These conditions made political campaigning difficult in Saskatchewan. A Liberal nomination meeting was held in Moose Jaw on 23 February 1974. Highway 42 north of Moose Jaw was closed, and much of the area was blocked with two-metre drifts of snow on that cold, stormy night.

The results of the ballot showed that Randy Devine had received 60 votes, Percy Lambert 154, and Colin Thatcher 236. Lambert was sure that the weather and road conditions had played a role in this outcome. Many of Thatcher's supporters lived near Moose Jaw, he reasoned, and so could get in to vote, while Lambert's supporters had been blocked by the weather.

There was no recount. Who knows how the history of the province might have changed if the weather hadn't (apparently) been working in Colin Thatcher's favour.

based on an anecdote in *A Canadian Tragedy*, by Maggie Siggins

The shallowest snow cover occurs in southern Saskatchewan, which reports average values of less than thirty centimetres. Snow cover over fifty centimetres can be expected for the boreal forest and for all of Manitoba, with the exception of the southwestern agricultural portion of the province.

Winter nights are bright when the moon is up. The reflective capability of fresh snow is high, and there is often enough light to cross-country ski or take a sleigh ride without the aid of artificial lights. It's a good thing, too, otherwise winter evenings might be unbearable. Even in the south, the

The Year of the Blue Snow

That winter has remained ever since, in the minds of all who went through it, as the true measure of catastrophe . . . after 1907 no one would seriously value those earlier disasters. The winter of 1906–07 was the real one, the year of the blue snow.

Wallace Stegner, *Wolf Willow*

Snowbank Suicide

One warm day in early spring, snowmobilers were enjoying the happy combination of mild weather and fine snow cover when they made a gruesome discovery near Landis, Saskatchewan: an arm protruding from a snowbank. A dead man's hiding place was being uncovered by the spring melt.

He was a stranger who had shown up one winter day, sitting in the bar until it closed for the night. The hotel owner offered him a room, but the stranger said it wasn't necessary. He walked to the edge of town and shot himself. Snowfall covered the body, and he wasn't found until spring.

The local people learned that the man had come from Saskatoon. He had recently quarrelled with his wife, and they had decided to separate. He immediately left for Edmonton. When his car stalled south of Landis, he hitch-hiked into town. There, despair must have set in as he sat drinking in the bar all day, brooding on his bad fortune. Then he made his fatal decision. His car sat on the shoulder of the road for several days before someone thought to tow it into Landis and start looking for the owner. By then, of course, he was buried beneath the snow.

longest nights can stretch to about sixteen hours. In the far north of the prairie provinces, the darkness lasts about eighteen hours at the winter solstice.

Most records for low daily, seasonal, and annual temperatures were set early in the history of the prairie climate stations, in the late 1890s and early 1900s. The winter of 1906–07 was one of the worst of these record-setting events. Following an extremely wet summer, winter started with light snow in the beginning of November. A three-day blizzard followed at mid-month, and was succeeded by a series of bad storms. By December, the range cattle were dying in huge numbers.

Wallace Stegner immortalized that winter in his book *Wolf Willow*. He describes the cowboys fighting to keep cattle alive in the Whitemud River Range in southwestern Saskatchewan, and losing the battle. That winter changed the life of the region. Cattle men who had ridden the ranges for years took a second look at the hordes of dead cattle that littered the prairie in the spring, the stench of putrefaction, and decided to retire. A few big ranches continued, but the rest of the region was turned over to homesteaders.

Dale Young waterskiing in February on the South Saskatchewan River. (Bernie Lukan)

One Bad Winter

The snow ploughs couldn't keep up during the winter of 1955-56 in Handel, Saskatchewan. They were on the go constantly, often getting the roads open by midnight only to have them blown in again by morning. On 12 December, a particularly brutal storm hit the area during school hours. The temperature dropped drastically. Visibility reached zero within an hour. Some students suffered frost bite before reaching safety. The railway tracks were blocked from Wilkie to Kelfield for most of the winter. The highways were also blocked, with snowdrifts well over two metres deep. Handel was completely cut off, except for those few people who had snowplanes or bombardiers—or those who still owned horses.

based on an account in *Handel New Horizons*

The Ice Man Skieth

One February an ardent water skier volunteered to help out a local water ski club in their promotions. Despite the warmth of his intentions, however, Dale Young found the water a trifle chilly that February day in 1973. He also found the floating ice something of a challenge as he skimmed and swerved across the surface of the South Saskatchewan River in Saskatoon. He was wearing a full wet suit, but that was scant protection against water temperature of 2°C, air temperature of -15.6°C, and a wind-chill of - 34°C. A good-sized crowd had gathered to watch him, including representatives of the media, and the hoar frost on the trees by the Bessborough Hotel provided a dramatic backdrop as he plunged into the frigid water. Ironically, he didn't lose his balance or hit a piece of ice; he fell because the boat's motor failed. He was immediately fished out of the river. His father rushed him home to a hot bath and a hot toddy. The water ski club raised enough funds from the stunt to build a ski jump on the river, and the only part of Dale's body to suffer frostbite was his heel. Even so, no one could accuse him of having cold feet.

Spring—The Green Wave

By March, temperatures are rising and the days are quickly growing longer. In most years, the snow cover is rapidly disappearing. But Arctic air masses can linger, and spring snowstorms can occur as late as the end of May in the grasslands, even later in the boreal forest. Snow banks linger into June in some years. In others, we get scorching temperatures early in the season, and April can feel like mid-July, with temperatures above 30°C. Spring is a confusing time on the prairies because it can feel more like a wicked combination of winter and summer, often on the same day, rather than a gentle transition into warmer weather. Spring brings new meaning to "out of the freezer and into the fire."

Spring is defined by climatologists as the months of March, April, and May. Astronomically, however, the first day of spring is 20 or 21 March, when the length of day and night are equal. Using temperature as a guide, the first day of spring occurs when the average daily temperature creeps above freezing. According to this definition, spring generally begins in the first part of April in places such as Winnipeg and Saskatoon, but arrives several days earlier in Calgary.

By mid-spring, average temperatures remain lower than -10°C in the northeast but have climbed above 5°C in the southwest. April daily high temperatures are greater than 10°C in the southwest, in Alberta and Saskatchewan, but they are still below freezing in northeastern Manitoba. Daily low temperatures range from below -15°C in the northeast to warmer than -3°C in the southwest.

Extreme temperatures in April are reminiscent of both winter and summer. Heat waves can bring temperatures above 35°C in the southwest, while cold spells have brought temperatures below -38°C in the north and -28°C in the south.

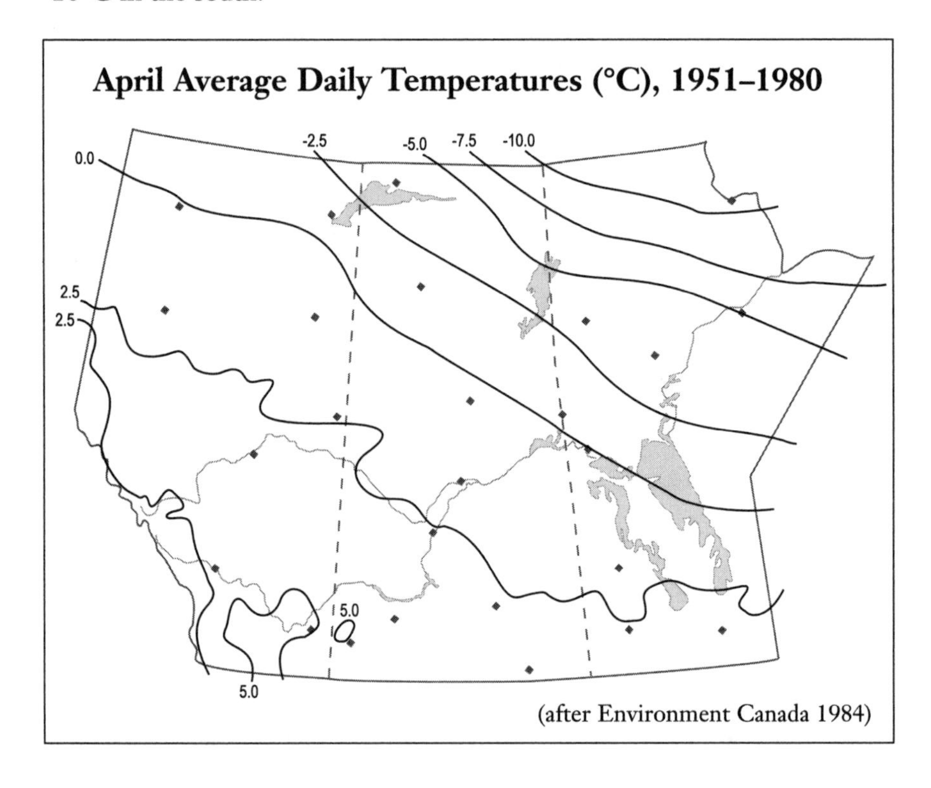

April Average Daily Temperatures (°C), 1951–1980

(after Environment Canada 1984)

Spring Blossoms

Crocuses herald the arrival of spring on the prairies. About ten per cent of prairie crocuses are blooming by 10 April, and most are purpling the hills and ditches by the 23rd. Aspens come into blossom a day or two later. The famous Saskatoon berry blossoms much later, in the middle of May, while the wily chokecherry waits to escape the May frosts. Others, such as brown-eyed Susan and yarrow, do not bloom until the latter part of June.

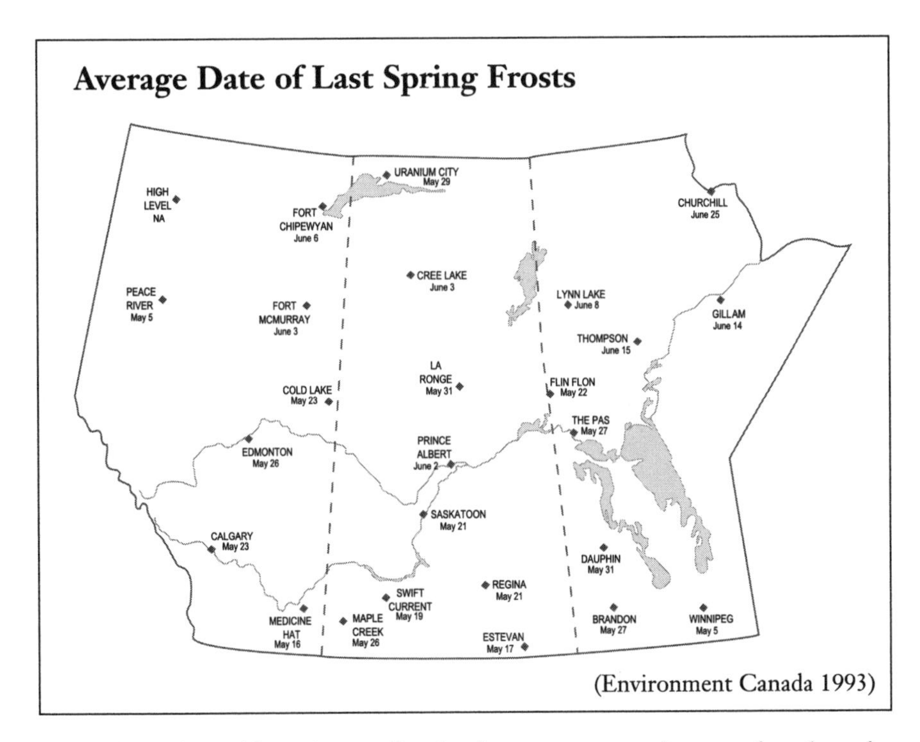

Plants have been blooming earlier in the season over the past decade, substantiating evidence of a warming trend on the prairies. Events in the life cycles of plants' lives are timed to respond mainly to heat accumulation in the first part of the season. Later in the season, the length of the day becomes a primary trigger.

So, when can prairie people start to think seriously about seeding gardens and crops in the spring? In general, the final spring frosts occur as early as mid-May in the southern prairies, or as late as the end of June in northeastern Manitoba. But as any farmer will tell you, frosts are highly variable, and gardeners should depend on frost-hardy plants with a short growing season.

Soil temperatures at ten- to fifty-centimetre depth range are usually just climbing above freezing at the end of March in the mid-prairie region. At depths of 150 to 300 centimetres, where gophers and water lines can be found, soil temperatures have fallen to their coldest levels in spring, about mid-March—just above 0°C at 150 centimetres, and 3°C to 4°C at 300 centimetres.

The Baby in the Ice Cave

y father and sister once rescued a baby from an ice cave. Dad had checked the cattle one day during the spring melt, and found a calf missing. He searched the hiding places the mothers usually used, checked the babysitter cow, even used his special calf call, but the baby was nowhere to be found. Then he noticed that the dog had a curious expression. It was looking down the lake bank into a tunnel that had been carved by the spring melt in the snow. He manoeuvred himself down to have a look into the dark cave, and there was the poor calf. The next challenge was to rescue her before she died. Dad was too large to fit into the tunnel himself, so he enlisted my sister's aid. Karen was just the right size to slither into the confined and slippery space. Dad fitted her with a harness and lowered her down. After that it was a relatively simple, if exhausting, task to pull the calf to safety. They took her directly to the house for revival. The calf not only survived, but it grew into a healthy cow that eventually gave birth to many healthy babies—none of which ever disappeared into ice caves. My sister and the dog certainly earned their keep that year.

Ah, Summer

Prairies swelter as mercury rises over 100°F (38°C) mark. . . . High temperatures, made more oppressive by hot winds, continued to hold the prairies in their grip during the weekend. . . . Seeking relief from the heat, people crowded lake resorts and swimming pools.

Saskatoon *Star Phoenix*, 6 July 1937

Summer on the prairies is not for the weak-hearted. We have many wonderful summer days, but we also have some severe ones. Some days the heat and wind combine to make you feel you're in a convection oven; when you're "done," you'll be as dry as a raisin.

July is usually the month with the highest temperatures, although the season often surprises us with cool spells. Average July temperatures on the prairies range from over 20°C south of Winnipeg to lower than 15°C in northeast Manitoba. Most of the boreal forest ranges from 15°C to 17°C, while the agricultural areas range from 17°C to 20°C.

The pattern of the isotherms is similar to the January patterns, in that they are angled from northwest to southeast. There the similarity ends.

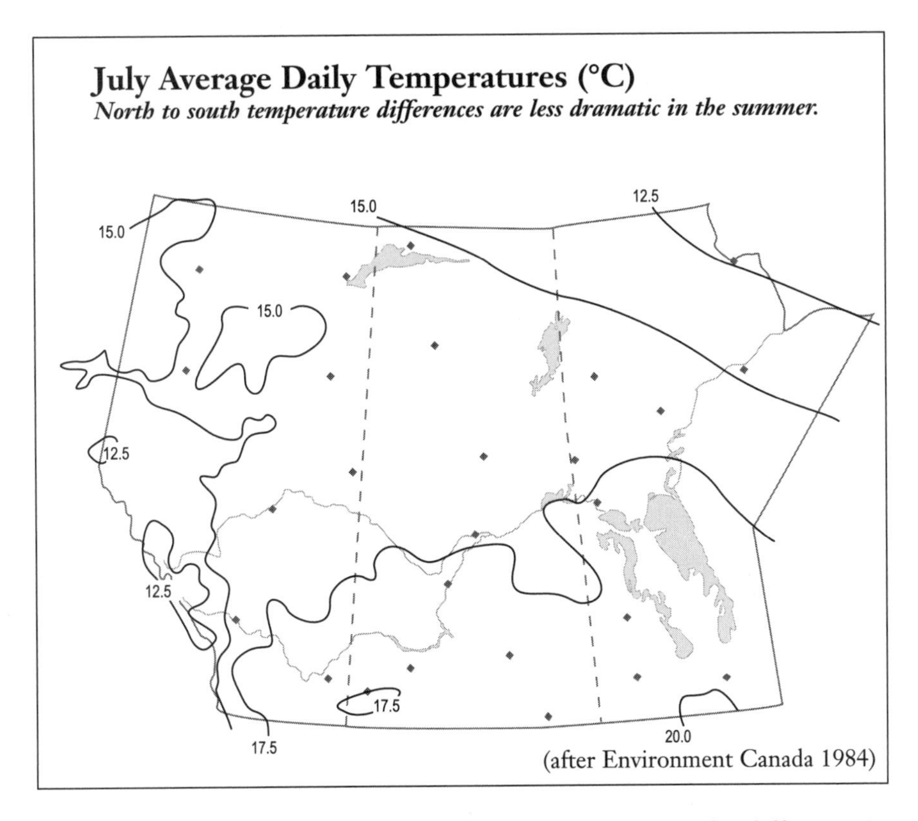

(after Environment Canada 1984)

The July isotherms are less regular and more spread out. The difference in average temperatures over the area is only about 5°C in the summer, but over 20°C in the winter. So the difference between the colder northeast and the warmer southwest is not as exaggerated in summer as it is in the winter. Consequently, summer vacationers can expect less difference in average temperatures in the north as compared to the conditions they're used to in the south.

Heat Wave Hazards

The highest temperature ever recorded in Canada was 45°C on 5 July 1937, at Midale and Yellowgrass in southeastern Saskatchewan. The southern prairies, especially southeastern Alberta and southwestern Saskatchewan, suffer the highest number of hot days on the prairies, averaging more than twenty per year. The number of hot days drops off rapidly toward the northern edge of the grasslands, and is less than five for most of the boreal forest.

When hot days follow one upon another, the effect is much worse than

A football player seeks icy relief from the scorching summer heat. (Gord Waldner)

a single day or afternoon of above-normal temperatures. Buildings become more difficult to cool. Streets and sidewalks accumulate the heat and seem to intensify it. Mercifully, summer heat waves are shorter than winter cold spells; most break in less than a week. And it's a dry heat on the prairies, unlike other parts of the country where heat is invariably accompanied by humidity. There is usually a wind on the prairie. And it is cooler at night.

Heat waves are almost as hazardous as winter cold spells. On average, eleven Canadians die every year from heat-related injuries, including heat cramps, heat exhaustion, and heat stroke. Heat stroke occurs when the body's thermo-regulatory system breaks down after prolonged exposure to excessive heat. Heat exhaustion results when blood volume is reduced through excessive loss of water or salt during sweating. Heat cramps are muscle spasms in the extremities, back, and abdomen of people who drink large amounts of water but fail to replace the salt lost from sweating.

Even more people die from conditions such as heart attacks and other problems that are made worse by heat stress. There is a strong correlation between higher temperatures and death from all causes. Early summer heat waves are especially hard on people and animals. Later in the season we're all more acclimatized to the warm weather.

Summer Breezes

Unlike winter wind-chill, summer winds can provide welcome cooling relief during hot spells. Another advantage of windy summer days is that insects are blown away. The strong, persistent winds also make sailing fun and make bikes with eighteen gears almost practical.

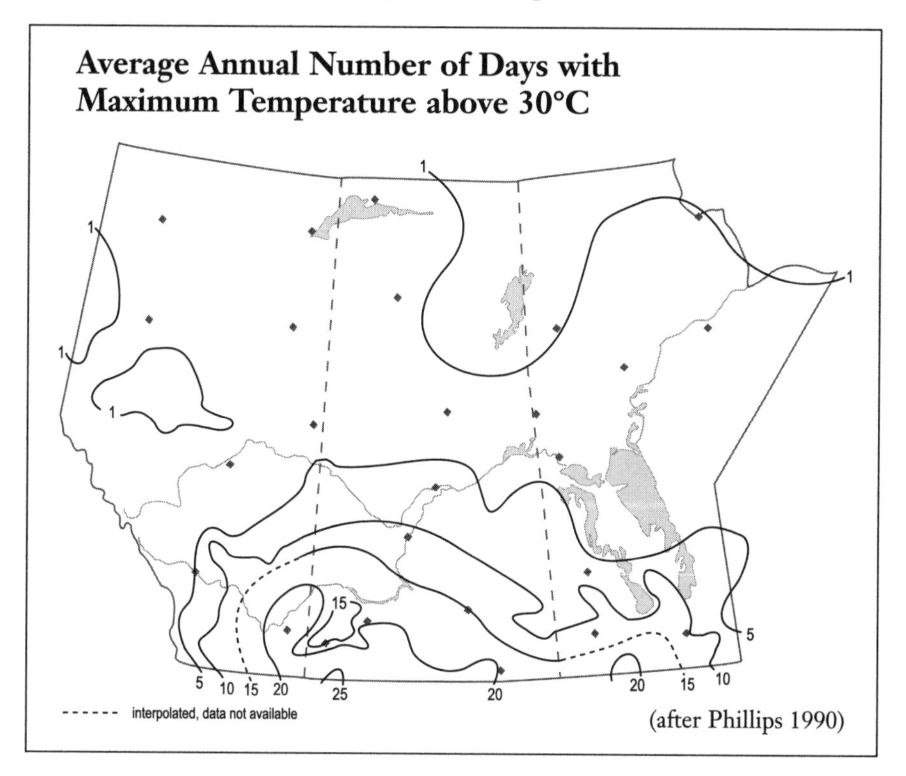

Average Annual Number of Days with Maximum Temperature above 30°C

(after Phillips 1990)

At twenty-four kilometres per hour, the mean wind speeds for July are highest in the Swift Current area in southwestern Saskatchewan. They are almost as high in the Churchill area of northeastern Manitoba. They drop to their lowest values—less than 7 kilometres per hour—in the boreal forest, especially in northwestern Alberta.

The Year Without a Summer

Even the coldest summers are better than the summer of 1816. In 1815, Mount Tambora erupted on the island of Sumbaw, Indonesia. An intense, sun-blocking dust veil formed high in the atmosphere, resulting in extremely low temperatures across the Northern Hemisphere. Because of these unusually low temperatures, 1816 is called the year without a summer.

Although little historical information about the climate of this time is available for the Canadian prairies, journals from Hudson Bay posts attest to the "little ice age," complete with daily temperature records. More evidence is available for the New England area, which was severely affected by cold waves during the summer of 1816. One storm which ended on 11 June left seven to fifteen centimetres of snow in northern New England. Cold weather continued to plague the area throughout the summer. Killing frosts struck on 9 July, and 21 and 30 August. All but the hardiest grains and vegetables were destroyed. If such outbreaks of cold were general across North America, one can imagine how severe conditions were farther north on prairies.

The conditions that brought snow to New England in June also brought severe conditions to the Hudson Bay region. The western Bay area escaped the full force of the weather, but records for the southeast indicate that continual frost and snow throughout the summer resulted in extremely poor vegetation growth compared to other years.

More recently, the summers of 1992 and 1993 were reminiscent of 1816. Except for people in British Columbia, Canadians suffered through an usually cold and wet summer in 1992. The prairies and Ontario also had to contend with hailstorms and freak snowstorms. Many social and economic activities were affected by the unseasonable cold, including agriculture, forestry, tourism, and road construction. Demands for power increased appreciably, and even beer sales dried up with the lack of appropriate quaffing weather. The costs of the cold summer approached half a billion dollars.

Costs of the Cold Summer Weather of 1992

Sector, Type of Cost	Cost or Effect
Agriculture	
Lost Production	$250–$350 million
Crop Insurance Indemnities—Premiums	$207 million
Forestry (cold winters mean little snow cover, and lead to tinder-dry springs)	
Forest Fire Suppression Cost (1992–1991)*	$15.1 million
Hectares Burnt (1992–1991)	510,456 ha
National Parks	
Park Entrance Revenue, Decline from Forecast (1992)	$728,550
Utilities	
Ontario Hydro net income, Decline from Forecast (1992)	$40 million
Hydro Quebec net income, Decline from Forecast (1992)	$0.4 million
Air Conditioners (2nd & 3rd Quarters)	
Residential Air Conditioners, Decline in Shipments (1992–1991)	7,665 units
Commercial Air Conditioners, Decline in Shipments (1992–1991)	5,458 units
Residential Air Conditioners, Decline in Sales Revenue (1992–1991)	$46 million
Road Construction	
Increase in Total Cost for Firms in Ontario	4.5-6.0%
Beer Sales (May through August)	
Decline in Summer Beer Sales (1992–1991)	56,404,900 litres
Decline in Sales Revenue (1992–1991)	$125,071,000
Total, Canada	$447,085,000 – $547,085,000

*Where 1992-1991 means "1992 values minus 1991 values"

(Herbert 1993)

Endless Summer Days

The prairie summer may be short, but it makes up for it with very long days. While people in the tropics are faced with darkness by about 6:00 PM, prairie people have the pleasant prospect of about four more hours of sunlight at the end of June in the agricultural area, and more farther north. We gain an extra half-day of summer every day for many days. Also, twilight

Rain has not dampened the spirits of these folks, though umbrellas are often inadequate protection against prairie weather. (Gord Waldner)

Extreme Heat Waves

Event and Date	Comments
Canada's worst heat wave, July 1936	For a week and a half, temperatures over 32°C prevailed from southern Saskatchewan to the Ottawa Valley. Crops were blackened. 780 people died.
Temperatures of 45°C recorded at Midale and Yellowgrass, Saskatchewan, 5 July 1937	Canada's record high temperature.
Prairies, June 1961	A record hot and dry month in a record drought year.
The 1980s	The warmest decade recorded in Canada and the world. Drought, dust storms, and hot spells were severe on the prairies.
Earth's average temperature for 1997	The earth's hottest year on record.

(Phillips 1990 and 1993, with additions by the author)

is a much more gradual event on the prairies than in the tropics. In fact, on the northern prairies it never really gets dark in the last part of June—you could fish or hike all night if you wanted to.

Summer Rain

Despite the dryness and the wind, summer is actually the wettest season on the prairies, where seventy to eighty per cent of the average annual precipitation falls as rain. June is usually the wettest month in the southern agricultural region, with July holding that distinction in the boreal forest.

As we have seen, the Palliser Triangle in southwest Saskatchewan and southeastern Alberta is the driest region of the Canadian prairies, with annual rainfall averaging less than 300 millimetres, which is much lower than the rate of evaporation. Averages increase as one moves north into the boreal forest, east of Winnipeg (over 400 millimetres), and west of Edmonton (over 400 millimetres).

The Icy Heat

It seems surprising that the heat of summer often brings frozen precipitation, but it's true that only the hottest weather produces the intense thunderstorms that turn into hail-makers. Hail grows in a thunderstorm or cumulo-nimbus cloud as frozen water droplets, dust, and other objects—even insects—collect layers of ice in the cloud. They travel the currents of the cloud until they become too large to be swept up by the updrafts. To reach the size of golf balls, hail stones must stay in the cloud for five to ten minutes.

Cedoux, Saskatchewan, has the dubious distinction of being the recipient of Canada's largest hail stone. It weighed 290 grams and measured 114 millimetres in diameter. In comparison, the heaviest hail stone in the world, which fell in the Guangxi region of China on 1 May 1986, weighed 5 kilograms.

Although hailstorms are menaces throughout the southern prairies, damaging hail is most frequent in the high plains of Alberta, a region west of a line running from Edmonton to Calgary. The area experiences, on average, more than five days a year with hailstorms. At more than three days per year on average, the hail hazard in southwestern Alberta and southern Saskatchewan is also high.

Average Annual Number of Days with Hail

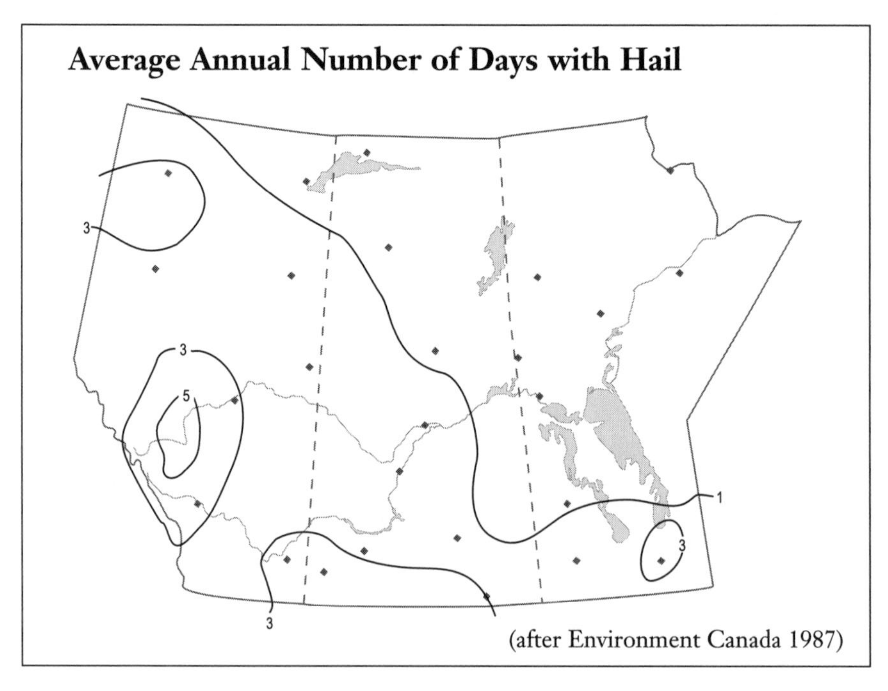

(after Environment Canada 1987)

Hail is a costly hazard to crops and property of all types. Hail storms can move rapidly, wreaking havoc along a path up to 150 kilometres long and three to twenty kilometres wide. Several storms may strike an area in a single day, and outbreaks may last two days or more. Crops can be completely destroyed in a matter of minutes. In "hailstorm alley" in Alberta, an average of three per cent of the total crop is wiped out by hail each year, and crop insurance payments for hail in Saskatchewan are greater than for drought in most years.

The urban hail hazard, too, has increased as cities expand. A devastating hail storm struck Calgary on 7 September 1991, causing about $350 million damage—a loss comparable to or higher than the damage to the entire prairie crop in some years. Calgary has experienced other heavy losses due to hail, as have Edmonton and Montreal.

Hail Storm Losses in the Canadian Prairies

Location	Date	Estimated Loss ($millions)	Comments
Calgary	7 September 1991	350	Greater damage than for the entire prairie crop in some years.
Moose Jaw	8 July 1989	9	
Edmonton	31 July 1987	100?	
Prince Albert	14 August 1982		The risk of hailstorms decreases as you move north, so this storm was rare.
Calgary	28 July 1981	120	This fifteen-minute hailstorm set a record for Canadian insurance claims for hail.
Western Prairies	23 July 1971	20	The hail swath was over 500 km long; severe weather lasted two days.
Edmonton	4 August 1969	17	The storm dropped some of the largest hail stones ever observed in the area.
Cedoux, SK	27 August 1973	10	The largest documented hailstone in Canada was dropped by this storm: 290 grams, 114 millimetres across.
Edmonton	10 July 1901		Most tin roofs were ruined and thousands of lights were broken. Hailstones measuring 80 millimetres and 140 grams fell.
Central Alberta	14 July 1953		Hailstones the size of golf balls damaged 1,800 km^2. Thousands of birds were killed.

(Data: Paul 1994, Phillips 1990)

Fall

Geese calling, combines growling into the night, leaves swirling, cattle lowing for their weanling calves, disappearing daylight, red sunsets on golden fields, crows gathering, risk of frost: the signs of fall sneak up on us quickly. Daytime highs in the thirties are soon displaced by temperatures of 5°C and 10°C. When we are faced with the fact that winter will soon be here, the warm fall days seem precious and few. The fresh, crisp air, the lack of bothersome insects, the colours of the trees and shrubs and fields as they transform themselves in their autumn colours render this brief time even more precious.

Fall begins with September, or the first day the average temperature drops below freezing, or the autumn equinox on 22 or 23 September. Climatologists worldwide define fall as September, October, and November. On the prairies, though, it seems a much shorter season than these three months would suggest. Spring and fall are the "sandwich" seasons between winter and spring, and often seem more like a mixture of these seasons than a gradual transition.

Fall is usually warmer than spring. Temperatures during the middle month, October, range from less than -2°C in the northeastern prairies to

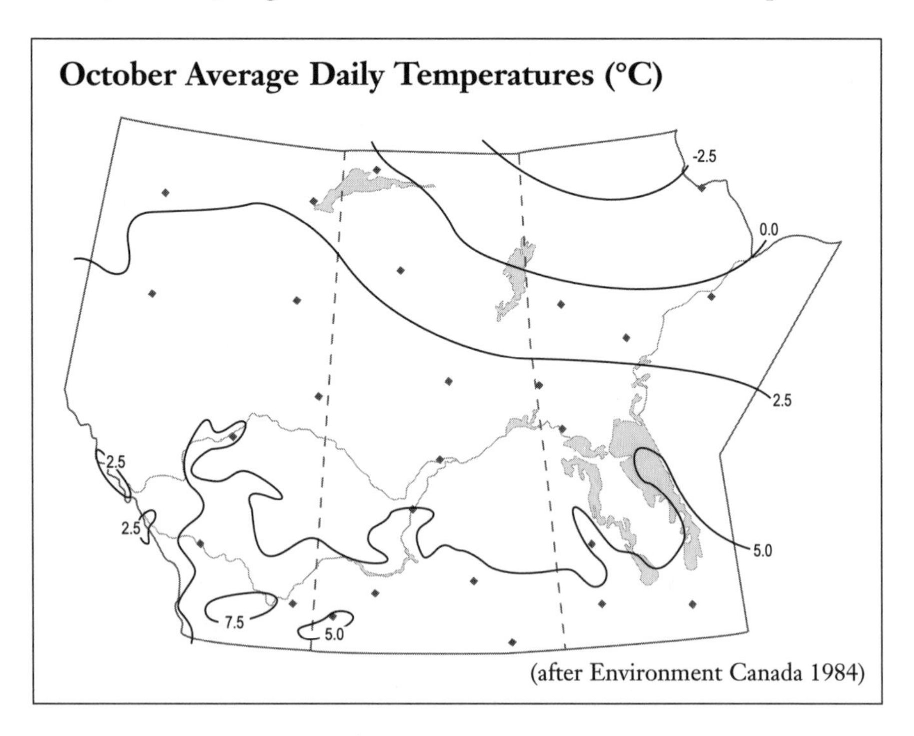

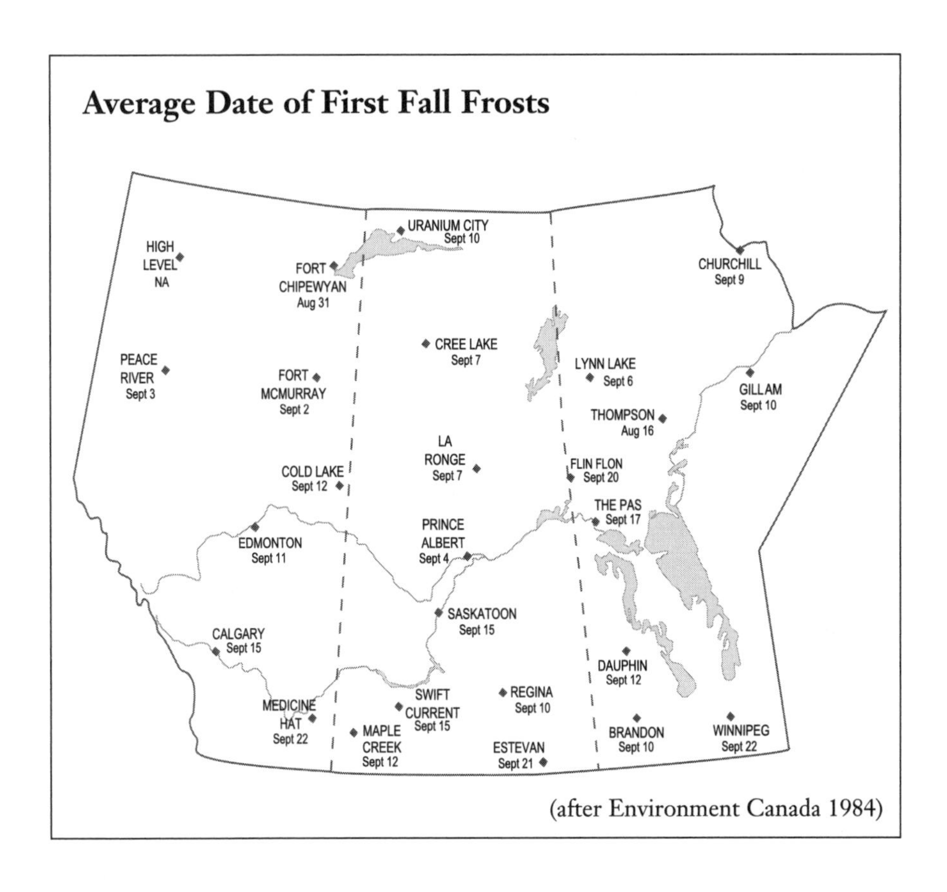

(after Environment Canada 1984)

higher than 8°C in southern Alberta. October's average highs are usually 2°C to 5°C above the mean, and the daily lows average 2°C to 7°C below the mean. Compared to mid-spring, mid-fall's average temperatures are about 8°C warmer in the northeast and 3°C warmer in southern Alberta. Winter keeps its grip on spring, as summer keeps its grip on fall.

This cool fall effect can be seen more clearly on graphs of the seasonal march of average temperatures for Winnipeg, Saskatoon, and Edmonton (page 42). October's mean temperature for these cities is very similar, converging on about 5°C. April's mean temperature, in contrast, is about 2°C or 3°C lower than October's.

Spring and fall average, or expected, temperatures are similar for these cities, even though their winter and summer temperatures diverge. Calgary has the most noticeably different temperature climate of these places. It is several degrees warmer during the winter and about 3°C cooler than the other major cities in the summer.

Everything is Relative

My climatology professor used to comment with amazement at the differences in how people would react and dress during the fall as compared to the spring. In the spring, at the first sign of warmth, parkas would be replaced by T-shirts and shorts. At the same temperature in the fall, in contrast, the students would be wearing parkas.

Humans and animals easily become acclimatized to hotter or colder temperatures. In the spring, conditions seem mild at values that would seem cold if we had experienced them in the fall, before we became acclimatized. Likewise, the relatively mild winters of the 1980s and early 1990s spoiled us for the long, cold winter of 1995–96, not to mention the frigid winters of the 1950s, or the death-dealing cold of the late 1890s and the turn of the century. Whether a winter seems warm or cold is more a matter of what we have become used to than the actual temperatures.

Winter's First Breath

Frost is a dreaded word for farmers and gardeners alike, especially if they have susceptible, unharvested crops. Frost strikes early, about mid-September, in many places across the agricultural prairies. Boreal forest locations can expect frost by the first part of September in the south, by the end of August farther north.

Areas affected by frost are highly variable from place to place. In the south, for example, a valley with a "frost hollow" may have a shorter frost-free season than a more favourably situated area farther north, such as a slope. Other conditions favouring frost include clear skies at night, dry evening air, calm or light winds, and dry soils.

Year-to-year changes are also considerable. One year may bring splendid harvest weather, with summer-like conditions lasting through the fall. Other years might bring snow in August, and weather conditions that force farmers to abandon the harvest, with the hope of being able to finish it in the spring. The latest fall frosts have been recorded late in October at many stations, while the earliest have been recorded in mid- to late August for stations farther north.

The first frost, of course, marks the end of the short prairie growing season. The length of the frost-free season determines not only what types of crops farmers and gardeners can grow, but the quality of the harvest as well. Many agricultural and garden crops are matched quite closely to the length of the growing season. Spring wheat requires 100 frost free days, oats 95, tomatoes 105. The length of the frost-free season averages only 120 days over much of the southern agricultural prairies, and less than 100 for the boreal forest region, dropping to lower than 80 in the far north.

Winter, spring, summer, fall: we complain about our climate, but we're certainly never bored with it. We don't have to travel to experience a wide range of weather. Rather, the weather comes to us, ranging from Arctic to tropical within a single year, or even hours.

References

Ball, G. 1989. The funnier side of weather: The Beaufort Wind Scale . . . somewhat revised. *Weather* 44(3):130.

Beaubien, E. 1993. Phenology: Using plants as biological weather instruments. *Climatic Perspectives* 15 (November):22–23.

Beaubien, E., and D. Johnson. n.d. Draft. *Plantwatch Alberta: A phenological approach to assessing the plant response to climate variability.*

Catchpole, A. J. W. 1985. Evidence from Hudson Bay region of severe cold in the summer of 1816. In *Climatic change in Canada: Critical periods in the Quaternary climatic history of northern North America*, edited by C.R. Harington. In *Syllogeous 55*. Ottawa: National Museums of Canada.

Environment Canada. 1993. *Canadian climate normals 1961 to 1990 on diskette*. Version 2.0E. Downsview, ON: Environment Canada.

Flysac, Larry. 1996. Personal Communication. Larry Flysac is a climate technologist at Environment Canada.

Environment Canada, Atmospheric Environment Service. 1984. Map Series 1: Temperature and precipitation, and Map Series 2: Precipitation. *Climatic atlas, Canada: A series of maps portraying Canada's climate*. Downsview, ON: Canadian Climate Program, Environment Canada, Atmospheric Environment Service.

______. 1987. Map Series 3: Pressure, humidity, cloud, visibility, and days with thunderstorms, hail, smoke/haze, fog, freezing precipitation, blowing snow, frost, and snow on the ground. Climatic atlas, Canada: A series of maps portraying Canada's climate. Downsview, ON: Canadian Climate Program, Environment Canada.

______. n.d. *Wind-chill*. Internal fact sheet. Saskatoon: Saskatchewan Environmental Services Centre, Environment Canada.

Etkin, D., and A. Maarouf. 1995. An overview of atmospheric natural hazards in Canada. In *Proceedings of a tri-lateral workshop on natural hazards, Sam Jakes Inn, Merrickville, Canada, Feb. 11–14, 1995*, edited by D. Etkin, 1-63 to 1-92. Downsview, ON: Environment Canada.

Handel New Horizons. 1990. *Handel notes and half notes*. Handel, SK: Handel New Horizons.

Harington, C. R., ed. 1985. Climatic change in Canada: Critical periods in the Quaternary climatic history of northern North America. In *Syllogeous 55*. Ottawa: National Museums of Canada.

Herbert, D. 1993. The cost of summer 1992 weather. Could be half a billion dollars? *Climatic Perspectives* 15 (November).

Lutgens, F. K. and E. J. Tarbuck. 1992. *The Atmosphere: An introduction to meteorology*. 5th ed. Englewood Cliffs, New Jersey: Prentice Hall.

McMillan, D. 1995. Claims are pouring in. Regina *Leader Post*, 29 August.

Paul, A. 1996. Personnel communications. Regina: University of Regina.

Phillips, D. 1990. *The climates of Canada*. Ottawa: Canadian Government Publishing Centre.

______. 1990. Weatherwise. *Canadian Geographic* (December/January).

______. 1995. How frostbite performs its misery. *Canadian Geographic* 115(1):20–22.

Russel, A. 1993. *The Canadian cowboy*. Toronto: McClelland & Stewart Inc.

Saskatoon *Star Phoenix*. 1937. Prairies swelter as mercury rises over 100 mark, 6 July.

Siggins, M. 1985. *A Canadian tragedy: JoAnn and Colin Thatcher: A story of love and hate*. Toronto: Seal Books.

Statistics Canada. 1994. *Human activity and the environment*. Ottawa: Statistics Canada.

Stegner, W. 1962. *Wolf Willow: A history, a story and a memory of the last plains frontier*. New York: Ballantine Books.

Wood, R. A., ed. 1996. *The weather almanac*. 7th ed. Detroit: Gale Research.

Young, M. 1996. The teenage guide to winter dressing. *Western People* (24 October):2.

Chapter Three

Living with Drought

Drought is an intrinsic and dramatic part of the prairie climate. Hardly a year goes by without some type of drought occurring somewhere in the prairie provinces. Droughts do occur in other parts of Canada, but the hazard is not as intense, widespread, or frequent as it is on the prairies.

Droughts, classified as major natural disasters, affect Canada and the world in disastrous and surprising ways. Although the prairie agricultural area is more famous for its droughts, even the northern boreal forest is susceptible to them, especially in Western Canada. Droughts plague the forest with low river and lake levels, creating ideal conditions for forest fires.

Prairie people have lived through some very dry times, including the classic droughts of the 1930s, 1961, and the 1980s. We have become drought experts; it's part of our character. Newspaper headlines concerning the drought of 1988 are evidence of our preoccupation and concern, and the themes headlined in the newspapers reflect this concern. Their diversity is also amazing, ranging from wind erosion to ducks, cattle, power generation, and politics.

Defining Drought

Drought affects disparate areas and activities differently. Each of us could probably write a unique definition according to our own perspective. A major challenge in drought investigation, then, is defining and measuring the phenomenon. A colleague of mine defines drought simply as a "worrisome lack of precipitation."

Many prairie people, especially in rural areas, have been faced with water shortages caused by droughts. What do you do when the water level in your well is sinking? when all that comes out of the tap is sludge? when your crops are withering in the heat? when your animals have little water and diminishing feed? Each of these concerns could be defined separately. But only scientists are worried about definitions when water runs short, so why bother? We bother because any scientific assessment of the nature of drought—including its frequency, intensity, and the area affected—requires an objective means of measuring it. A definition of drought and its effects is the first step in this process, and a necessary prelude to policies for response.

Meteorologists frequently refer to drought as a long-term lack of precipitation. An agrologist may consider drought to be a period during which soil moisture is insufficient to support crops. Some plants are drought tolerant, while others may be considered "water hogs." Droughts can be specific to certain plants. A drought that might affect a susceptible crop such as canola, for instance, may not be so harmful to wheat.

A hydrological drought may mean a prolonged period of unusually low surface runoff, and low water levels in shallow wells. An agricultural-crop drought may occur during a year with flooding problems on rivers. During the 1988 drought, for example, Manitoba Water Resources personnel drove through a dust storm to reach a flooded area they were monitoring.

Ground water is vital to the prairies, especially in rural areas. Ground water responds differently to drought, depending on the level of the water-bearing strata and the type of drought. A short drought affects shallow wells more quickly than deeper wells, which may in fact remain unaffected. Shallow wells, however, often recharge more quickly with rainfall and snow melt. A drought may last off and on for several years before we see changes in deeper wells.

Selected Newspaper Articles about the 1987 and 1988 Droughts

Date	Paper/Magazine	Theme/Headline
June 22, 1987	Biggar *Independent*	Crop report/no precipitation/soil erosion
Nov. 19	*Western Producer*	Old Wives Lake dry
Dec. 19	*Leader Post*	No snow cover
Feb. 17, 1988	*Star Phoenix*	B.C. drought (5th year)
Mar. 21	*Leader Post*	Canadian water diverted to US?
Mar. 24	*Western Producer*	World weather "bad"
Mar. 26	*Star Phoenix*	Water levels low
Mar. 29	*Leader Post*	Estevan dry
Apr. 7	*Western Producer*	Soil erosion
Apr. 23	*Star Phoenix*	Costs of drought
May 5	*Western Producer*	Ranchers unload cattle
		3rd driest winter in Lethbridge
May 11	*Leader Post*	Drought vs. waterfowl

continued on page 81

May 31	*Leader Post*	Grassland fire near Elbow
June 1		Drought stops political squabbles
June 3	*Star Phoenix*	Next century weather forecasts
June 13	*Alberta Report*	Costs for one farmer
June 14	*Globe & Mail*	Projected crop losses
June 23	*Star Phoenix*	Ground water vs drought
June 25		1980s drier than 1930s
June 29	*Leader Post*	Increase in electricity costs
	Melfort *Journal*	Field erosion
June 30	*Western Producer*	Weed/pesticides/heat
July 11	*Alberta Report*	Unemployment vs drought
July 19	Shaunavon *Standard*	Shaunavon has a crop!!
Sept. 30	*Leader Post*	Livestock payments (map)
Oct.	*Ag News*	Record amount of water assistance
Oct. 5	Estevan *Mercury*	Drought payments to cattle owners
Oct. 6	*Western Producer*	Manitoba vs 2 year drought
		Farmers can learn to make do with lack of water
Oct. 13	*Western Producer*	Drought aid in limbo
Oct. 27		No fall moisture
Nov. 21	*Alberta News*	Political aid for farmers
Jan. 5, 1989	*Western Producer*	Sask. economy survived drought
Feb. 9		World temperature in 1988 highest
Feb. 21	*Leader Post*	Snow cover sporadic
Feb. 27	*Sask. Farm Life*	Crop drought assistance
	Leader Post	World drought?
Mar. 9		Drought helped increase nitrogen
Mar. 16	*Western Producer*	Grain prices
		Sask. Water's water relief program
Mar. 22	Estevan *Mercury*	Drought zones defined
Mar. 30	*Western Producer*	Weyburn terminal handles less grain
Apr. 1	Winnipeg *Free Press*	Farmers plough dry wetlands

Measuring Drought

Measuring drought is almost as challenging as defining it. Drought is a complex combination of several factors, including precipitation, evaporation, transpiration, and water use. Evaporation and transpiration, in turn, depend on factors such as temperature, humidity, and wind speed.

The measurement of droughts is important for several reasons. Scientists need numeric descriptions in order to investigate the frequency and intensity of drought patterns. These numeric descriptors are called

Drought is a Prairie "Drinking Problem"

One of the worst "drinking" problems in Canada is the lack of good-quality drinking water on the prairies. Drought is the driving force behind this problem. A water-quality project at the Saskatchewan Research Council began because of the generally poor-quality and limited sources of rural drinking water on the prairies. Poor water quality can have adverse health effects. Livestock do poorly and can die without decent drinking water.

"drought indices." If we can't measure drought, we can't find out how often, how bad, how long, or where it occurs. Drought assistance and monitoring programs require measurements to be objective and effective. People who manage resources such as reservoirs, crops, forests, and pastures require specific information about droughts in order to make appropriate decisions. This is especially important on the drought-prone prairies.

Many different methods are used to characterize the nature, severity, and extent (some would say the "ruthlessness") of drought. Precipitation variability is one indicator of an area's susceptibility to drought. The coefficient of variation (CV), an indicator of variability, is one of the best measures of precipitation variability. Higher CVs indicate greater precipitation variations, lower reliability of precipitation, and an increased tendency toward drought.

Umbrellas Are Not Designed for the Prairies

I grew up on a farm near Biggar, Saskatchewan, in one of the driest areas on the prairies. When I was about five, I got an umbrella for my birthday. I waited and waited for rain. When it finally came, the wind blew the rain sideways beneath the umbrella, and then it blew the umbrella itself to pieces. I rarely carry an umbrella now, and often forget to take one on trips to rainier locations (i.e., almost everywhere). The umbrella is a poor tool for coping with rain on the prairies.

World Distribution of Coefficient of Variation of Annual Precipitation (percent)

Where in the world is the precipitation most variable?

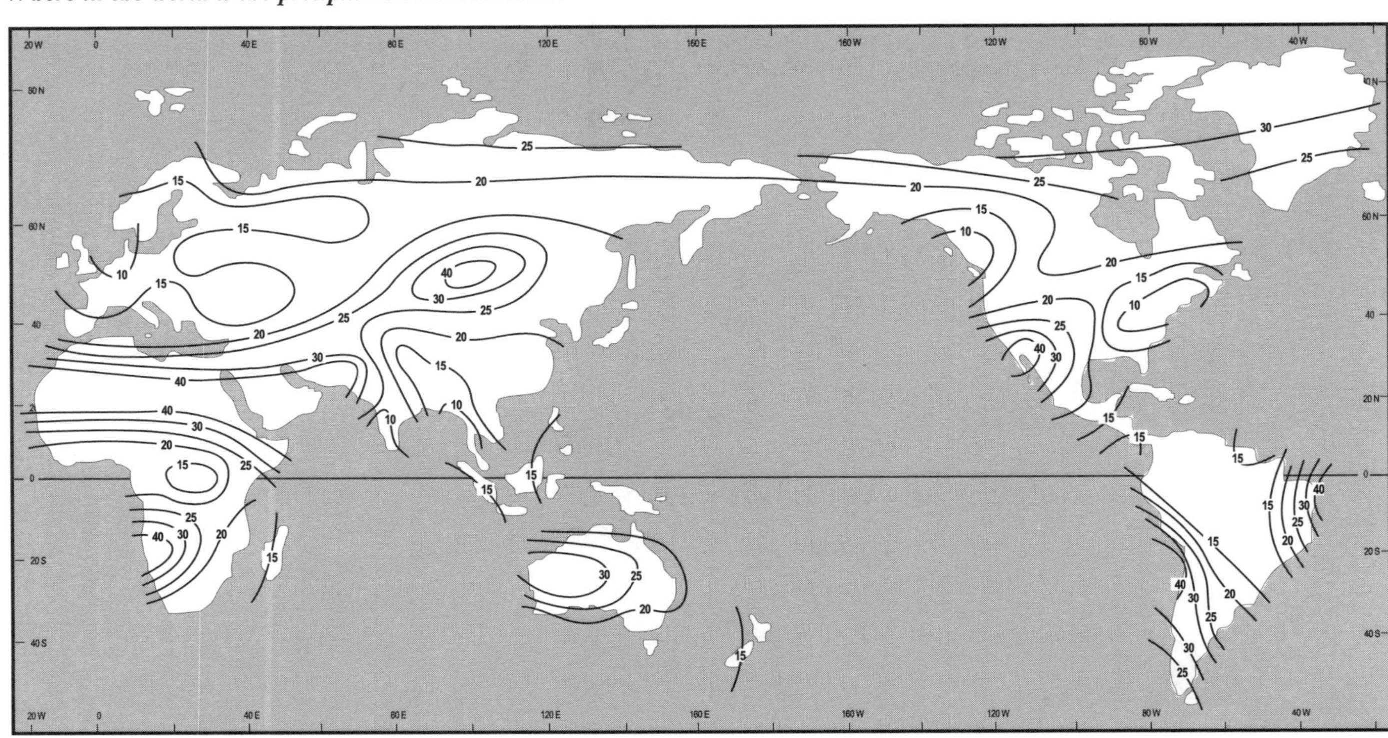

(after Barry and Chorley 1976)

The Hydro-illogical Cycle

The idea of the hydrological cycle is moderately well known. Water evaporates from the earth and rises into the atmosphere as vapour, where it cools, condenses, and forms rain. The rain falls to the earth, and the cycle begins anew.

The hydro-illogical cycle is not so well known. I first came across the concept while attending a conference in 1986. One of the presenters, Don Wilhite, used the idea as a way of explaining how planning for droughts occurs (or doesn't occur). Preparation should begin well before a drought begins, especially in areas that are prone to drought, and especially when other gains could be made by such activities. Drought planning, in fact, should be ongoing.

Planning and response, however, tend to be haphazard. And after the drought is over, instead of using the time to plan for the next one (which will be inevitable, although we don't know exactly where and when), we tend to forget about drought altogether. That is the "hydro-illogical" cycle.

Some areas have boringly similar amounts of rainfall from one year to the next. Coastal British Columbia tends to be one of them, with a CV of less than ten percent. Higher CV values are found in the arid regions of the world, including the Sahara and the Kalahari deserts of Africa, and the southwestern United States, which have CVs in the thirty- to forty-percent range. With values exceeding twenty percent, the prairie provinces have the second-highest values in Canada, second only to the high Arctic, which has values greater than thirty percent.

Precipitation reliability, as measured by the CV, is poor in the prairie provinces as a whole. Some are worse off than others. The area of best growing-season precipitation reliability is around Edmonton, which has a CV of about twenty percent. The worst area, once again, is in the Palliser Triangle, which has a CV of more than forty percent.

Precipitation variability also changes with the seasons, exacerbating drought problems on the prairies. The variability is greatest during the summer months when the various needs and uses for water for agriculture and recreation are highest. This is a serious and unfortunate mismatch. It means that the water supply is most sporadic and uncertain when the need is greatest. If the peak were to shift to winter, drought problems would decrease dramatically. On the other hand, if precipitation were to become even more erratic, drought problems would likely increase.

This year-to-year variability makes planning difficult. We can go for years without serious water shortages in a particular place, and become complacent about conservation. We might even be fooled into thinking that our water worries are over. Then another drought sneaks up, and there go our precious water supplies again. And there goes the forest, up in smoke again. So guarding and conserving our water supplies is critical on the prairies. We must be vigilant about conservation, pollution prevention, and demand management. What it really boils down to is the wise use of resources.

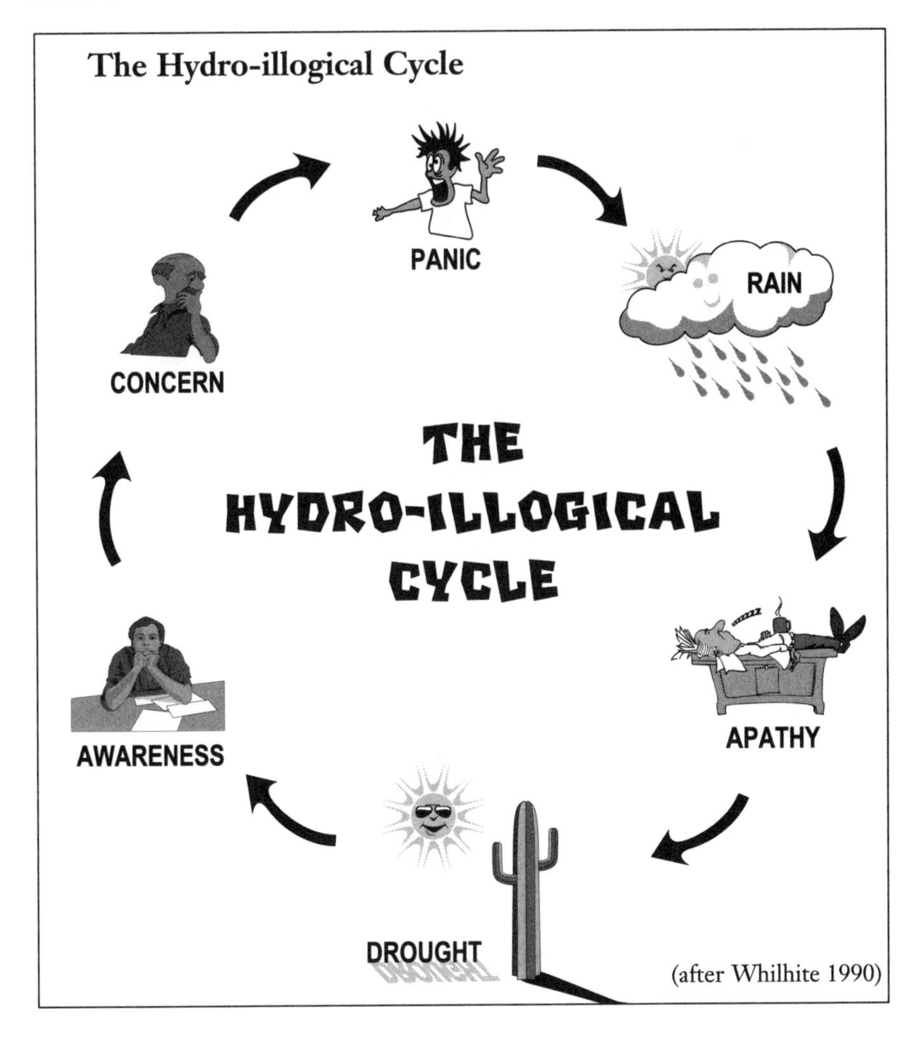

Place to Place and Time to Time

Prairie precipitation is not only variable from month to month, season to season, and year to year, but summer rainfall is also notoriously different from place to place. One field may be deluged while another remains bone dry. One part of even a small city may literally be filling with water, while another part gasps in the dust.

This is why prairie people often ask their neighbours, "How much rain did you get?" They're not just making polite conversation; they're curious to track the different rainfall patterns. These patterns can stretch the trust of insurance adjustors, too. But when farmers separated by only a few kilometres declare losses from opposite causes (flood and drought), their claims are usually legitimate.

The Palmer Drought Severity Index (PDSI)—a sort of drought "measuring stick"—is used extensively in North America to study and monitor drought conditions. The PDSI considers many factors, including precipitation, evaporation, transpiration, antecedent soil moisture, and runoff, and is useful for examining long-term, year-to-year drought occurrences. Many years between 1908 and 1993 brought dry to drought conditions (PDSI less than -2) to southern Saskatchewan. In contrast, few years have had wetter conditions (PDSI greater than 2).

Long-term drought patterns emphasize the year-to-year variability of

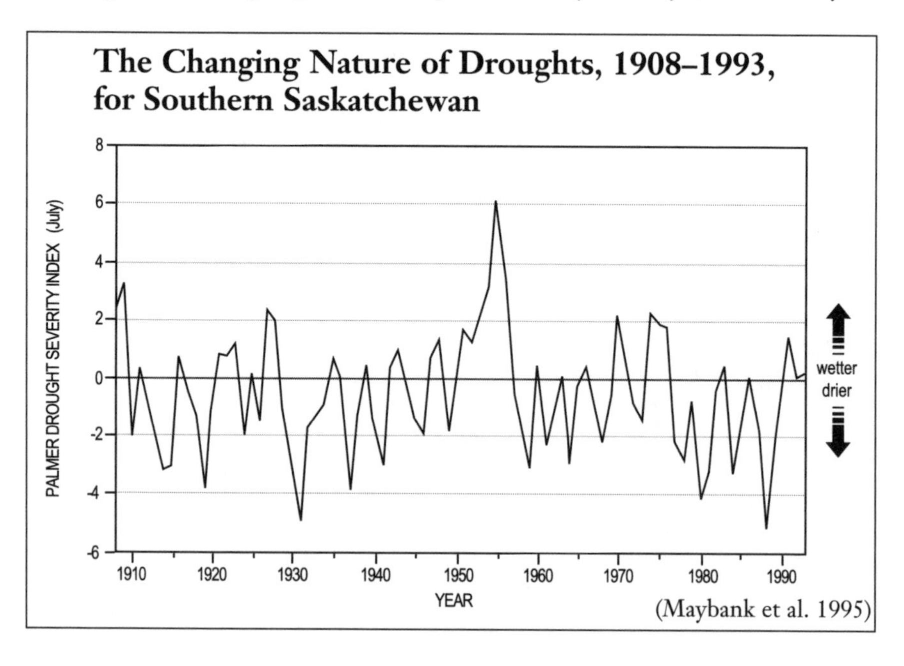

(Maybank et al. 1995)

drought. But trends can be plotted through this seemingly sporadic pattern. If you drew a line through the overall trend for the period 1974 to 1989, it would point downward, indicating a trend toward more droughts. In contrast, an earlier trend during the 1940s and 1950s was upward, indicating fewer droughts and more wet periods.

The Effects of Drought

Droughts have a major impact on the Canadian economy, environment, and society. Many activities related to agriculture, forestry, water resources, waterfowl, fisheries, recreation, energy, tourism, and transportation can be adversely affected by drought. It is also thought to contribute to heat-related health problems, even deaths.

Drought years on the prairies can be measured and compared by their effects. Certain droughts can affect one sector severely (agriculture, for example) but not another (such as water supply management). The most severe and prolonged droughts, however, such as occurred in 1987–88, often affect many sectors severely. A mild drought may only affect summer pastures, but a nasty one may affect sectors beyond the agricultural as its effects ripple through the economy. Such a drought is disruptive and costly.

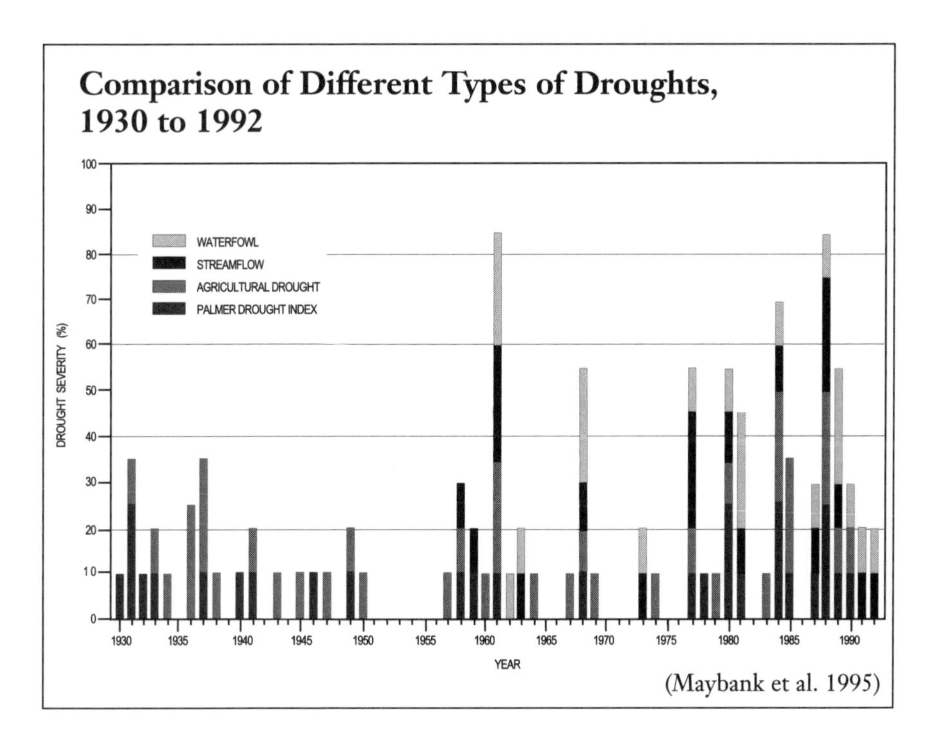

Comparison of Different Types of Droughts, 1930 to 1992

(Maybank et al. 1995)

Drought has its bright side, though: brilliant, sunny weather; larger beaches as lake levels drop; fewer mosquitoes; and more work for forest fire fighters. It also provides ideal conditions for outdoor activities, from construction and hay making to open-air concerts.

Drought Watch

Unlike most weather hazards, droughts do not have a swift, attention-grabbing progression. Rather, they sneak up through the seasons, catching us by surprise. By the time the crops start to wither and well levels drop, it may be too late to take precautions. It is important, therefore, to monitor for droughts and to have well-developed, tested water conservation and other drought fighting methods ready much of the time.

When a "budding" drought is detected, mitigation plans of various types—ranging from improved forest-fire watches to grassland management and soil conservation—must be initiated or enhanced to minimize destructive effects or maximize any benefits. Just as we have weather watches and warnings, we need to be alert to and prepared for droughts. Only this will reduce our vulnerability and alleviate the economic and social costs of droughts.

Water and soil conservation is a continuous requirement for drought-prone areas. Prairie people have developed and applied water conservation techniques for years. But we're still learning as nature throws new surprises at us.

Do Droughts Remember One Another?

The cause of droughts is not clear. They are usually related to persistent high-pressure systems which, in turn, are linked to the blocking of westerly winds by patterns of atmospheric circulation, the Rocky Mountains, or by combinations of such factors. Climate patterns in far-away places such as the Pacific Ocean also affect the prairie climate.

The position of the prairies, in the centre of a continent far from any ocean, is also a contributing factor. By the time air masses reach us, their moisture has often been wrung out by various processes, and they have none left to share. If there were a large inland sea reaching into Canada as the Mediterranean reaches into Europe, droughts on the prairies would be rare.

One interesting theory holds that droughts "feed on" or "beget" other

A drought victim—parched, cracked earth. (Elaine Wheaton)

droughts. For example, dry spring conditions favour the formation of high-pressure systems. Dry conditions also mean less local water vapour as a source for rain. This perpetuates the dry conditions for a season or longer. Dry soil conditions, then, appear to prolong and amplify droughts. This effect is aptly termed "persistence" or "memory" in the system.

Moist soils and growing plants, on the other hand, give off water vapour. If water vapour is being released into the atmosphere in this way over a large area, it provides a good local source for future rain if other conditions are appropriate. But if the soil is dry, the vegetation sparse, and water bodies minimal or nonexistent, the area has to depend solely on water vapour from other regions for its rainfall. This increases both the probability of drought and the lifetime of an existing drought. The circle is self-perpetuating, as less rainfall leads to fewer plants and drying soils and water bodies. The dust and dried-up sloughs and sparse vegetation of 1988 were not only caused by, but increased the lifetime and intensity of the drought.

Another link in the chain of cause and effect is the colour of the earth's surface, which affects the planet's ability to reflect and absorb sunlight. Bright surfaces, such as fresh snow or white sand, reflect much of the light that strikes them, so they absorb less, and less heat energy is trapped in the lower atmosphere. This results in the cooling and sinking of the air mass above, which, in turn, results in little or no rainfall. Dark, moist soils and a blanket of healthy plants, on the other hand, absorb much of the light that strikes them, resulting in a warm, rising air mass—ideal conditions to produce precipitation, if there is enough water vapour in the air.

Another contributory cause to climatic anomalies such as drought is a process termed "teleconnections," in which conditions in distant places are thought to be related to local conditions on the prairies. Certain patterns of surface temperatures in the northern Pacific Ocean, for instance, are associated with summer dry periods on the prairies. The weather pattern associated with the 1988 drought was a strong, high-pressure system stalled over the American midwest, displacing the jet stream and corresponding stormy weather systems north of their usual positions. These major shifts in the atmospheric circulation system were thought to play a large role in the drought, which was the result of more than simply dry surface conditions.

Infamous Droughts

Each prairie drought is unique, differing widely in terms of intensity, timing, area, duration, causes, and effects. The most notorious and devastating prairie droughts occurring in this century were the classic droughts of the 1930s and 1980s, and the single drought year of 1961. Other droughts have been noteworthy, but none has been as famous as the "Dirty Thirties" and the equally "Dirty Eighties." The 1961 drought was extremely severe—crop drought conditions prevailed over most of agricultural Manitoba and Saskatchewan, spreading into southeastern Alberta—but it was limited to 1961. Few areas in either 1960 or 1962 experienced a drought. Back-to-back, yearly droughts are much more difficult to recover from, both economically and socially.

Although the worst agricultural droughts occurred in 1936–37, 1961, 1984–85, and 1988, crop yield droughts affecting parts of the prairies occurred in at least thirty-two other years between 1900 and 1993: 1910, 1914, 1917, 1918, 1919, 1920, 1924, 1929, 1931, 1933, 1934, 1938, 1941, 1943, 1945, 1947, 1949, 1950, 1957, 1958, 1960, 1964, 1967, 1968, 1969, 1974, 1977, 1979, 1980, 1983, 1989, 1990. If nothing else, this clearly shows that drought on the prairies is common. Droughts as bad as the worst year of the 1930s occur, on average, once in twenty years.

Boreal forest droughts are less common than droughts on the prairies, and less noticeable because of the sparse population and climate stations in

A summer snowman framed by lilacs, 1 June 1982. This unconventional summer fun resulted from lingering snowbanks left behind by a widespread prairie snowstorm from 29 to 30 May. (Clara Moch)

Ranchers work through all kinds of weather, even Alberta snowstorms. (John Perret/Light Line Photo)

Farmyards were submerged by the creation of the Red River "Sea" during the great flood of 1997 in Manitoba. The spring flood level was the highest recorded this century. (Gerry Kopelow/First Light)

Storm clouds bring localized rain to a canola field. (Wayne Shiels/Lone Pine Photo)

What is it? Answer: A salt storm. Old Wives Lake in Saskatchewan dried out during the severe drought of 1988. As a result, salt-dust storms were frequent unwelcome visitors to the area that summer. (John Perret/Light Line Photo)

A dangerous sky rudely interrupts the peacefulness of a rural church. (John Perret/Light Line Photo)

Horses are not intimidated by towering castles of cumulus clouds that appeared in the late afternoon on a hot, humid day. (Adrian Cutler)

The semi-arid climate makes grass the king of the vegetation types in the southern prairies. (Clarence W. Norris/Lone Pine Photo)

What is it? Answer: Virga are cloud streaks caused by rain or snow falling from clouds and evaporating before they reach the ground. (Ottmar Bierwagen/Canada In Stock)

Lens-shaped clouds in waves over large round hay bales emphasize spring's colours near Pincher Creek, Alberta. (Darwin Wiggett)

Fair-weather, fluffy cumulus clouds are not yet threats to this flax crop.
(Clarence W. Norris/Lone Pine Photo)

Storm clouds form over Edmonton, Alberta. (John Luckhurst/GDL)

A tornado advances too close to the Baker farm, twelve kilometres south of Saskatoon, Saskatchewan, 27 June 1990. The tornado edged past the South Corman Park School just as buses were arriving to take students home. The students and teachers hid in the school until the weather calmed down. (Gordon Fehr)

Suitably timed and adequate rainfall, marked here by a rainbow, are necessary to produce grain to fill these elevators. (Ron Richardson/Here and There Photography)

A dust storm, c. 1988, billows over Minnedosa, Manitoba. (Brian Milne/First Light)

An intense rain slashes through shasta daisies. Prairie plants must be tough to survive.
(Clarence W. Norris/Lone Pine Photo)

An autumn scene in Alberta's Elk Island National Park. (John Luckhurst/GDL)

A prairie sunset travels down the tracks instead of a train. (Wayne Shiels/Lone Pine Photo)

Fall's golden hills and their shadows seem to roll on forever. (John Perret/Light Line Photo)

Buildings in a farmyard fade into a snowstorm. (John Perret/Light Line Photo)

Summer twilight lasts long into the evening. (Dale Young)

Crocuses struggle to bloom in the snow. (John Perret/Light Line Photo)

Winter hoar frost and bright sun turn a tree grove into a magical place. (John Perret/Light Line Photo)

Sunrise's horizontal light brings vivid colour revealing the depth and character of this cloud system over the Crowsnest Pass, Alberta. (Darwin Wiggett)

the north. Even so, the droughts of the 1980s also affected the boreal forest, and continued into 1994-95.

Droughts have occurred across the prairies since the last ice age, when the climate became semi-arid to sub-humid, but weather records for most climate stations usually go back less than 100 years. Other methods—historic records, and the study of tree rings and lake sediments—have been used to plot the recurrence of drought over longer periods. Most of these studies have been in and for the United States, but droughts over large areas often extend into Canada.

Major droughts on the great plains during the previous two centuries were clustered in the late 1750s, the early 1820s, the early 1860s, and the mid-1890s. Historical records mention drought in 1820 and 1868 in the Red River Settlement of what is now Manitoba. There are several references to droughts in the 1890s, but records are too scant for a comprehensive picture.

About 247,000 people left the prairies from 1931 to 1941. (Saskatchewan Archives Board/R-A4287)

The Dirty Thirties

As the thirties advanced, conditions on the prairies became progressively worse. Grasshoppers, drought, weeds, wind, and dust were all menaces to our continued livelihood. Crops were practically choked out. There was insufficient feed for animals. Gardens were poor. Our milk cow died (just before freshening) due to lack of feed. My mother cried because she had been looking forward to having fresh milk for us. As time went on, it became impossible to make payments on land and other machinery. Those bright tomorrows my parents had always talked about now went by the wayside. Everywhere, there was hopelessness. People began moving out. Many went north, and other places where they could find hope and a future. We were among the people that abandoned all we had hoped for in our prairie home. In October 1937 we left for the north.

based on an account in *Handel New Horizons*

Who Stole the Floor?

During the 1930s, dust storms were frequent and violent. Houses were not sealed as well as they are now, and the dust drifted in and out of the house. Glasses were routinely turned upside down as the table was being set (a habit that persists in many prairie kitchens) because, by the time the meal was ready, a layer of dust would have fallen in the glasses.

My father and his brothers were sleeping out in the bunkhouse during a particularly dusty summer. His story is that, one morning, they trooped in for breakfast and saw that the floor was gone! They raised the alarm while looking around for other lost articles. They thought that burglars had come during the night and stolen the floor. This sounded far-fetched to me. How could anyone steal a floor? Why would anyone want to? But kids have good imaginations. My grandmother set their wild suppositions to rest with a broom. The floor wasn't gone, but hidden under the layers of dust that had blown in during the night.

The Dirty Thirties

A wheat-yield-based drought indicator showed that 1929 to 1938 was the worst long-term drought in the fifty-two-year period from 1928 to 1980. The drought year of 1929 set the stage for the 1930s. The next year was not considered a "wheat drought" year, but moderately dry conditions prevailed in some areas. Drought hit again in 1931, affecting south-central Alberta, central agricultural Saskatchewan, and a portion of southwestern Manitoba. There was adequate moisture again in 1935, but crops were devastated by an epidemic of wheat-stem rust.

Despite the hardships of the first half of the decade, it was the period 1936 to 1938 that ensured that the drought would become infamous. Reported wheat yields were lower than for any other three-year period from 1921 to 1974. The droughts of 1936, 1937, and 1938 were the second, third, and fourth worst droughts, respectively, during the period from 1928 to 1980.

Statistics can be as dry as drought itself, but these statistics are staggering. The drought and economic depression combined to devastate the

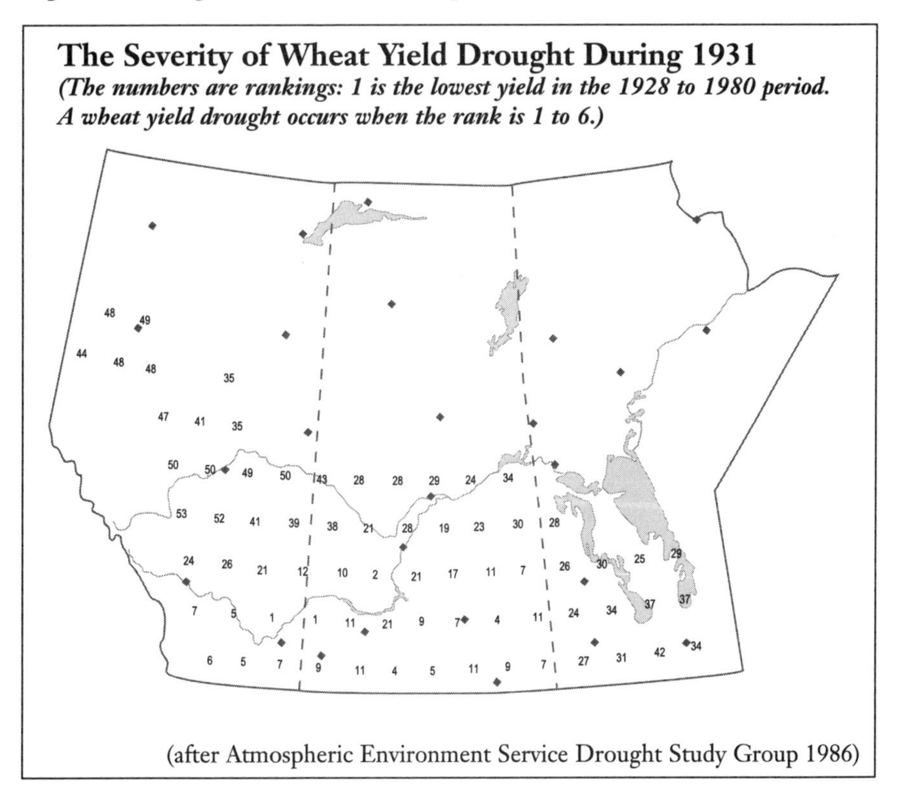

The Severity of Wheat Yield Drought During 1931
(The numbers are rankings: 1 is the lowest yield in the 1928 to 1980 period. A wheat yield drought occurs when the rank is 1 to 6.)

(after Atmospheric Environment Service Drought Study Group 1986)

Extreme June Temperatures of the 1988 Drought

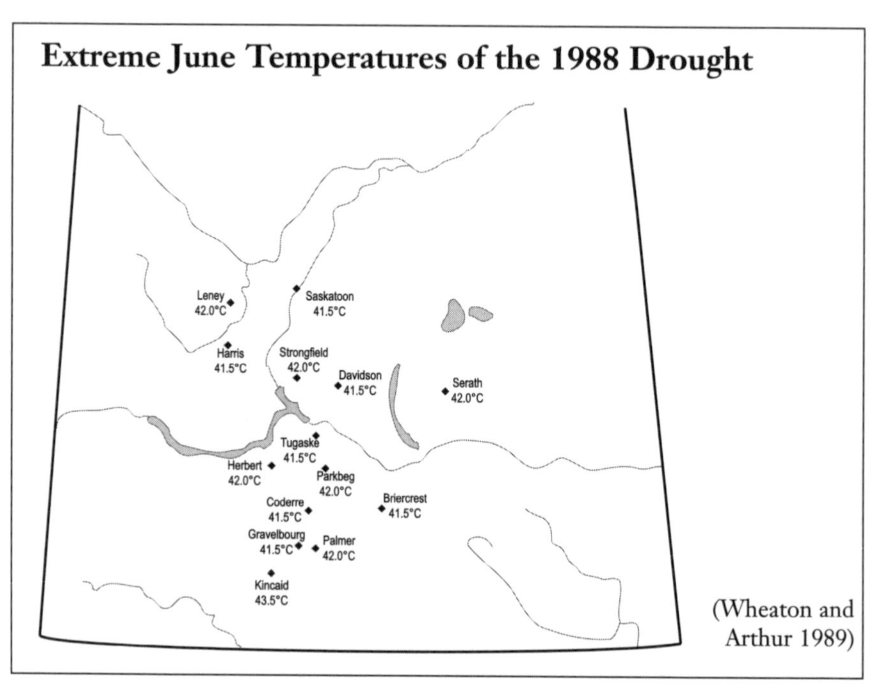

(Wheaton and Arthur 1989)

Drought Patterns in the Canadian Prairies, June 1988

Each drought has its own characteristics.

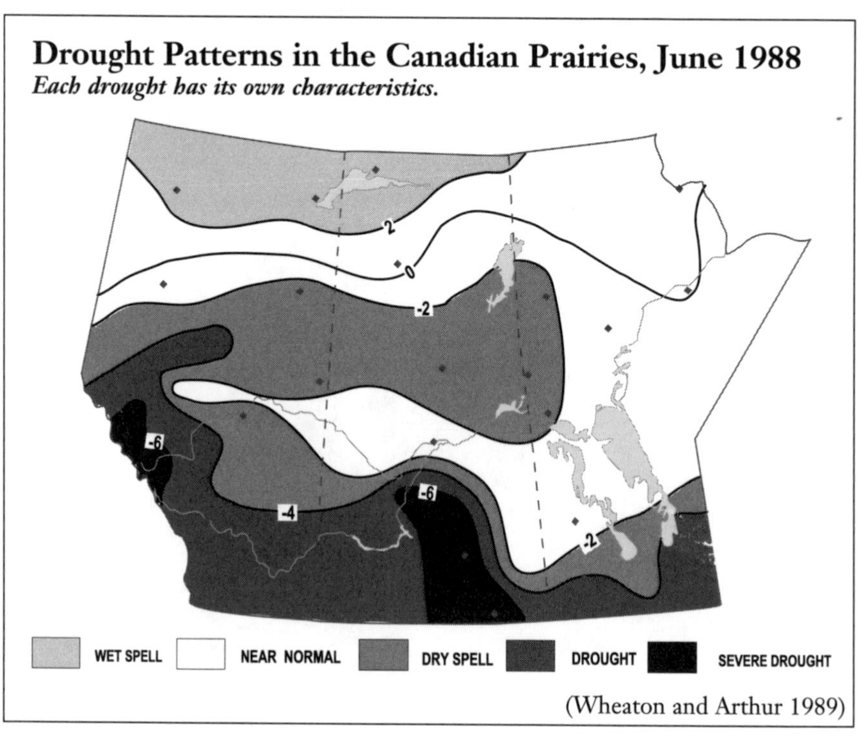

(Wheaton and Arthur 1989)

people, the environment, and the economy. Farm abandonment in the thirties was unprecedented. Saskatchewan reported 4,907 farms abandoned in 1926, 5,193 in 1931, and 12,831 in 1936. Malnutrition was endemic, especially among children. By 1936 the wheat belt had changed into the dust bowl.

Severe environmental effects included the "black blizzards" or dust storms that could blow for days. They sandblasted crops and gardens, destroying what was already weak from lack of water. Thousands of hectares of land blew out of control; tonnes of valuable topsoil were lost. And of course the dust storms added to the misery and discomfort of people and animals.

The Dirty Eighties

The dry fall of 1987 was a harbinger of one of the worst droughts in North American history. The drought of 1988 is remembered for its severity, its persistence, and the extent of its reach across the great plains of North America. Above-average temperatures and below-average precipitation were reported for the winter of 1987–88 for all three prairie provinces. By May and June 1988, nature's blast furnace was turned on high, and temperatures were soaring. Even night-time temperatures were unbearable. Usually, the pastures are a wonderful green

The Dirty Eighties

The eighties brought about numerous difficulties for farmers in the district. Drought, low farm prices, high production costs, combined with high interest rates caused a serious struggle, especially for young farmers. In addition, they were plagued with grasshoppers and flea beetles. Moisture levels in the eighties were equivalent to those of the thirties. There was a trend to milder winters with less snowfall. Changes in farming methods, fertilization, herbicides, and pesticides enabled farmers to produce reasonable yields, but they also had to rely on crop insurance, grain stabilization programs, and Canadian Crop Drought Assistance Program payments, as well as production loans for grain and cattle to help allay farm costs.

based on an account by John Eppich *Handel New Horizons*

blanket by May; in May 1988 the fields were brown and dry, and the grass crunched underfoot when you stepped on it.

June 1988 was the hottest ever recorded for most stations, with temperatures exceeding 40°C for several days in a row. Several stations broke records for extreme highs. The record high was almost broken as Kincaid reached 43.5°C.

Humans can withstand a wide range of climatic conditions. We have survived temperatures ranging from 50° to -60°C. Most people, however, function best in the "comfort zone," a few degrees above or below the optimum annual average of about 10°C. The threshold temperature of this comfort zone is relative, however; it is lower in cooler climates and higher in warmer climates. People in cooler climates are therefore more vulnerable to heat waves than those in warmer climates.

Heat waves can increase or exacerbate health problems, sometimes resulting in death. The summer of 1988 affected many people across North America. Twenty-eight people in Saskatchewan were hospitalized in 1987-88 because of problems related to the effects of heat and light. There was one fatality. Eighty-eight were hospitalized for similar problems in 1988-89. Many others were affected by respiratory problems complicated

Travelling tumbleweeds trapped in a ditch during the 1988 drought. (Elaine Wheaton)

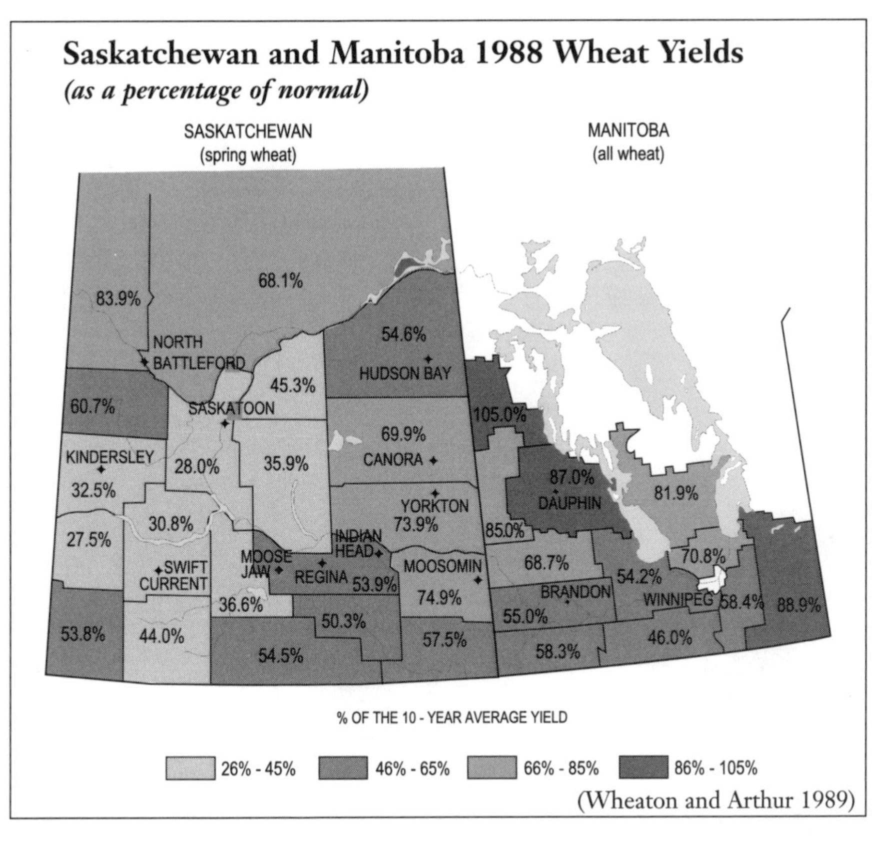

Saskatchewan and Manitoba 1988 Wheat Yields
(as a percentage of normal)

(Wheaton and Arthur 1989)

by poor air quality because of wind erosion and larger dust storms.

As one would expect, agriculture was one of the hardest hit sectors of the economy during the drought. Saskatchewan was the province hardest hit—a common pattern during drought years. Many farmers were unable to produce the eight bushels per acre necessary to meet harvesting costs. Over half the crop districts had yields less than fifty percent of the ten-year average. Only one district had yields that exceeded seventy-five percent. Compared to the previous year, wheat production in 1988 dropped by fifty-four percent in Saskatchewan, nearly thirty-nine percent in Manitoba, but only about five percent in Alberta.

The only plants that appeared to do well were weeds. Thistles sprang up everywhere. Mounds of them filled dugouts and ditches. They lined the fences, turning them into hedges for long distances. Kochia, another drought tolerant weed, grew as tall as humans.

Livestock producers were also hit hard. The southern Canadian prairies and northern plains of the United States experienced the worst conditions

Droughts increase the risk of raging forest fires. (Saskatchewan Environment and Resource Management)

for livestock because of dry pastures, poor hay and feed production, and shortages of water. Many cattle were moved to northern areas where pasture and water conditions were better.

Dust storms and wind erosion were extremely bad in 1988, which had not only some of the driest weather on record, but some of the dirtiest as well. The number of dust storms met or exceeded record numbers at several locations. Kindersley, in west-central agricultural Saskatchewan, experienced the maximum, at fifteen. The Winnipeg area came second.

Most water supplies on the prairies, even in the north, were severely affected. Dugouts, wells, as well as municipal and hydroelectric reservoirs shrank or went completely dry. Stream flows across the agricultural prairies measured less than half their average volumes in many places.

Drought can foster disease and insect outbreaks among trees, but its most noticeable and newsworthy effect is fire. Except for the previous record year, 1987, the number of forest fires in Saskatchewan, at 980, was higher in 1988 than in any other year since 1918. Fortunately—and surprisingly—the 81,000 hectares of forest lands destroyed by fire in 1988 was less than the long-term average of 130,000 per year. This was likely because of extensive and effective fire-fighting activities. Fire-fighting costs were high in both 1987 and 1988, at $33.9 and $31.8 million, respectively.

Ducks and other waterfowl were also affected by the drought. They and their preferred habitat—sloughs—became scarce during the 1980s as compared to normal. A record of the number of sloughs is kept on the prairies because they are important to waterfowl. By July 1988, the number of sloughs counted in Saskatchewan and Manitoba was below half of normal. Waterfowl were going farther north to find habitat. Even there, their numbers diminished because of crowding and increased incidents of disease thought to be related to dry conditions. On the brighter side, waterfowl damage to crops was much lower than usual that year.

Recurrent droughts are a part of the prairie climate. Some areas experience drought almost every year. But the consequences of the 1988 drought were severe, numerous, and wide ranging. The timing was particularly bad, striking agriculturally dependent economies already weakened by years of low prices.

"Was the drought of the 1980s as bad as the 1930s?"

It's a common question, and a difficult one. As with most difficult questions about complicated issues, the answer is yes and no. The 1980s droughts, especially the drought of 1988, are thought to have been more widespread than those of the 1930s. Precipitation totals over the ten-year periods 1929 to 1938 and 1979 to 1988 were almost identical, but the eighties were more notable for the heat waves. Temperatures in the 1980s were much higher than those in the 1930s.

The thirties also had different winters. Winters during much of the 1980s were mild, with little snow cover—so little, in fact, that winter dust storms were reported in 1988. The 1930s and 1980s droughts were similar in intensity, but had different characteristics, including the more severe winters of the 1930s and the more severe hot spells of the 1980s. The latter were also hard on the environment, the economy, and the people, but because of drought programs and policies and improved land management, the hardships were not as severe as were those in the 1930s. This is a lesson for dealing with future droughts: adequate planning and support can pay off in decreased hardship.

Predicting Droughts

With better models, computers, and data, weather forecasts are slowly improving (by about one percent per year), even in the prairie region, which is one of the toughest areas to forecast. Seventy-two-hour forecasts are now as accurate as thirty-six-hour forecasts were in 1957. Accuracy decreases with forecast length, however, so three-day forecasts are accurate seventy percent of the time while four-day forecasts are accurate only sixty-two percent of the time.

Presently, there are no satisfactory methods of predicting climate in the middle latitudes for monthly or seasonal periods. Drought prediction requires a sophisticated knowledge of the patterns of climatic fluctuations through time and by area. It also requires an understanding of the causes involved. The complexity of the global climate system makes this a very difficult task.

The global climate is characterized by complex interactions among the land surfaces, oceans, atmosphere, and outer space. It is not possible, at present, to predict the arrival of the next drought. It is only possible to predict, with absolute certainty, that it will come, and it will be followed by others. Global warming will possibly bring more frequent and more intense droughts.

References

Atmospheric Environment Drought Study Group. 1986. *An applied climatology of drought in the prairie provinces.* Downsview, ON: Atmospheric Environment Service.

Barry, R.G. and R. J. Chorley. 1976. *Atmosphere, weather and climate.* 3rd ed. London: Methuen.

Berry, M., and G.D.V. William. 1985. Thirties drought on the prairies—How unique was it? In *Climatic change in Canada: Critical periods in the Quaternary climatic history of northern North America*, edited by C. R. Harington. In *Syllogeous 55*. Ottawa: National Museums of Canada.

Berton, P. 1990. *The great depression.* Toronto: Penguin Books.

Gray, J. H. 1978. *Men against the desert.* Saskatoon: Western Producer Prairie Books.

Jones, K.H. 1991. Drought on the prairies. In *Symposium on the impacts of climatic change and variability on the Great Plains, 11–13 September 1990, Calgary, Alberta*, edited by G. Wall. Waterloo: University of Waterloo.

Maybank, J., B. Bonsal, K. Jones, R. Lawford, E.G. O'Brien, E.A. Ripley, and E. Wheaton. 1995. Drought as a natural disaster. *Atmospheric-Ocean* 33(2):195–222.

O'Brien, E. G. 1994. Draft. Drought in Canada. Submitted to the *Canadian national report for the international decade for natural disaster reduction.* Regina: Prairie Farm Rehabilitation Authority.

Phillips, D. W. 1990. *The climate of Canada.* Ottawa: Supply and Services Canada.

______. 1993. *The day Niagara Falls ran dry.* Toronto: Key Porter Books.

Wheaton, E. E., and L. M Arthur, eds. 1989. *Environmental and economic impacts of the 1988 drought: With emphasis on Saskatchewan and Manitoba, Vol. 1.* Saskatoon: Saskatchewan Research Council.

Wilhite, Don, ed. 1990. *Drought assessment, management and planning: Theory and case studies.* Boston: Kluwer Academic Publisher.

Chapter Four

The Outlook for 2040 A.D.

Scientific and technical advances have changed the way we do many things. They have improved how we monitor the climate and how we understand and use climate information. We now know that our future climate will likely be different from our current climate. Climates have changed in the past, with interesting consequences. They are changing now, and will change in the future.

Swift Current, Saskatchewan: 2040

It is April, and farmers are finished seeding a full month ahead of the seeding dates of forty-five years ago. Range lands and gardens are turning green a month earlier. Summer has been erratically growing longer over the past few decades, and many people, especially farmers, have been adjusting their activities accordingly.

Swift Current citizens have been noticing that the local climate, and others around the world, are becoming more unusual at a faster pace. Years ago it was only the old-timers who remembered different weather patterns from an earlier time. Now even teenagers have stories about colder springs and longer winters. The climate has been growing hotter, year after year. There were occasional cool summers in the early 2000s, but even they were not as cool as the summers of the 1990s. Climatic changes had been sneaking up on people, but now they are accelerating, and people are remarking on the differences of just two or three years back. The changes are becoming more obvious to everyone, not just to climatologists.

May 2040 is much like late June forty years ago, with daily temperatures in the upper 20s and crops already maturing. July and August are almost unbearably hot. Daily highs in the upper 30s are no longer setting records, as they had in the 1980s, but are common. As winter space-heating demands decrease, sales of air conditioners continue to soar.

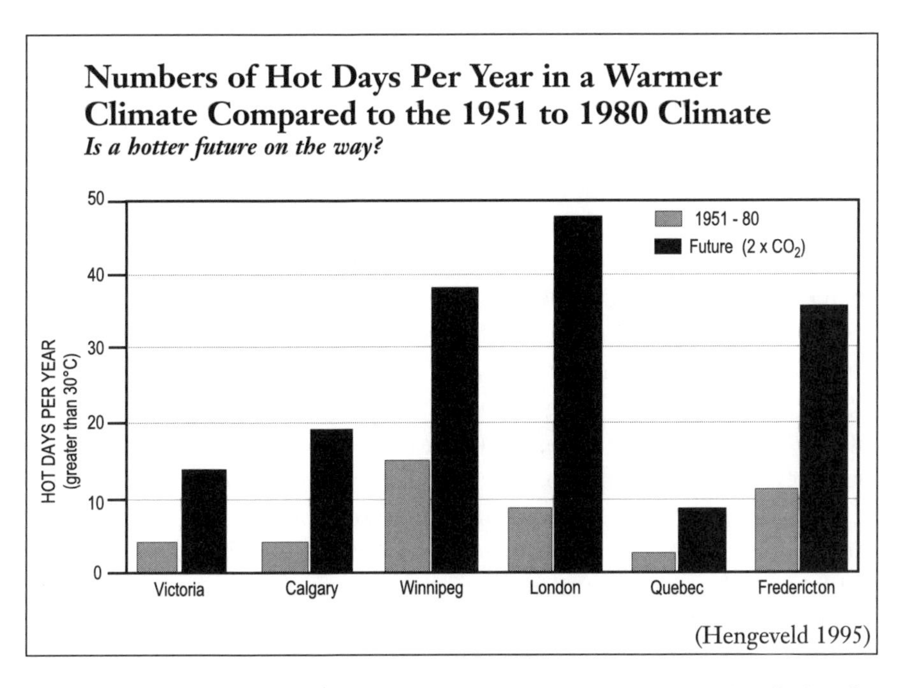

Air pollution reminiscent of industrial cities in the east has finally hit the prairies. In places such as Swift Current, the smog is exacerbated by the hot summer days and the abundance of sunshine. People have been unable to give up their cars, even though the taxes on carbon-based products such as gasoline are soaring. Vehicle pollution is largely responsible for the severe smog, while the more frequent and intense heat waves are especially hard on the old and the very young, and especially those without air conditioners. Respiratory and other types of environmentally related illnesses and deaths are increasing. Smog advisories are common, and more and more city dwellers are spending the summer at ranches and lakes to avoid the urban smog and heat.

The smoke from forest and grass fires sweeping over the prairies aggravates the effects of the smog. Decreased air quality means that more people with allergies have to stay indoors. Adequate visibility for travelling is of increasing concern. The numbers of forest fires have increased steadily with the rising summer temperatures and drought. There are few old-growth forests left. Replanting projects depend on the drought- and temperature-tolerant varieties of trees developed in the late 1990s and early 2000s.

Increases in the number of hurricanes were first reported during the

1980s and 1990s, and the 2030s set new records in the numbers, severities, and losses from hurricanes. Catastrophic windstorms across the globe have climbed from eight in the 1960s to fourteen in the 1970s, twenty-nine in the 1980s, forty in the 1990s, sixty in the 2000s, a hundred in the 2010s, and up to two hundred in the 2030s. Insurance agencies were among the first to pay attention to this trend as they found themselves, as early as the 1990s, with spiralling payouts for major storms and their damages. They changed their policies accordingly.

Climatic change has wreaked havoc on many of the world's cities. Heat waves in increasing numbers and intensity in megalopolises such as Chicago and New York were killing record numbers of people even in the late 1990s. At the same time, urban areas were flooding as sea-levels rose and rivers overspilled their banks during the intense rainfalls. Urban populations were hard hit by increasing water and food shortages, as well as the northward migration of tropical pests and diseases. Many people were heading north to cooler climates.

The population of Swift Current has tripled in the past thirty years. Refugees from climatic change effects stream onto the prairies not only from coastal North America, but from places as far flung as the Maldives,

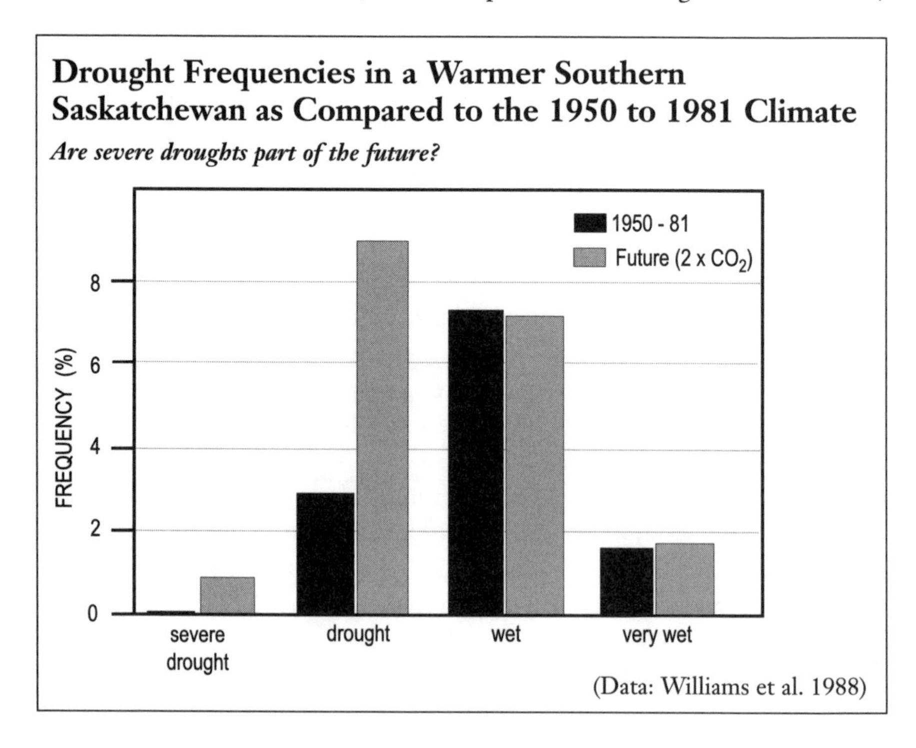

Drought Frequencies in a Warmer Southern Saskatchewan as Compared to the 1950 to 1981 Climate

Are severe droughts part of the future?

(Data: Williams et al. 1988)

Bangladesh, and the Seychelles. Once arrived, the refugees find themselves confronting climatic changes as widespread and alarming as the ones they left behind.

Climatologists predicted in the late 1980s that global warming would bring droughts of increasing frequency and severity to areas such as the southern plains. Climatic belts were advancing northward year by year. By

Street signs in a lake? A St. Germaine, Manitoba, resident uses an unusual mode of travel during the 1997 Red River flood. (Kevin Frayer, Canadian Press, 25 April 1997)

2040, Swift Current is getting hit twice as often by severe droughts. With increased knowledge and improved management, farmers are able to adapt to these changes by seeding earlier, enhancing soil and water conservation, planting drought-tolerant crops, and by relying on meteorologists and climatologists for information to make decisions that match the changing environment.

Some crops, however, have been less successful in the new climate. Canola is one such, and farmers have turned to more drought-tolerant oilseeds such as sunola. New forages that can tolerate midsummer droughts and take advantage of extended spring and fall seasons are also being used, and more fall-seeded crops are being planted to take advantage of the changing seasons.

By the 2010s, successive years of intense midsummer droughts had increased the number of dust storms. Black blizzards rolled across the land, stealing precious topsoil and dumping it into ditches, wetlands, and neighbouring provinces. Satellite images showed the storms even extending into the Atlantic Ocean. Farmers applied practices that were eventually able to counter climate-accelerated erosion by wind and water. But in the meantime, much topsoil had been lost, especially during the severe storms of the period 2011–15, years also marked by production losses and environmental and health problems.

Farmers have been able to sow crops requiring more frost-free days as the growing season has extended. This gives them the double advantage of diversification and higher yields. But because farmers continue to "push the envelope," frost continues to be a problem. Because of warmer winter periods, in fact, de-hardening of fall-seeded crops, trees, and other plants occurs more frequently.

Unfortunately, weeds, insects, and diseases have adapted more quickly to the changing climate than the crops and the farmers have. The warm winter of 1994–95 was a warning about the hazards of a warming winter. Grain beetles enjoyed the relatively balmy weather and caused serious problems in stored grain. The numbers of and problems with grain beetles intensified throughout the early 2000s. Such problems had decreased by the 2020s with improved adaptation technology, but the transition time has been difficult and losses have been large.

Storm Warnings

By 2030, midsummer droughts were often preceded by spring floods, because of an increase in snowfall in the Rocky Mountains and the intense spring rains. The summer of 1995 had provided a glimpse into the future, with floods, droughts, and forest fires striking simultaneously at different points across Alberta, Saskatchewan, and Manitoba. Climatologists had issued warnings of these coming conditions, but their reports were ignored by both decision-makers and the media.

Vastly improved water conservation efforts in and around Swift Current mean that the flood excess can be used to balance the water deficit brought on by midsummer drought. Illegal drainage of wetlands, however, continues to exacerbate the flooding problem on downstream lands. The floods are turning rural road maintenance into a nightmare, but many roads are at least passable now, thanks to the invention of special "mud kicker" vehicles by innovative prairie farmers.

Severe storms were common on the prairies throughout the twentieth century. By 2010 the most threatening tornado and dust storm belts in the midwestern United States had shifted northward into the southern plains of Canada. Storms had increased in both number and intensity, and the consequent damage was terrible. Even weather-wise prairie people could not always prepare for or protect themselves against these record-breaking storms, which were now occurring year after year.

In 2013, a devastating tornado even more severe than the killer of 1912 hit Regina, resulting in even more fatalities and greater damage. Improvements in forecasting and communications saved many lives and much property. Even so, forty lives were lost in the twister, twelve more than the previous record set a century earlier.

Shrinking Winters

People in Swift Current are accustomed to warmer winters now, and complain bitterly if the temperature drops to -20°C. Early projections indicated that winter climates would undergo the greatest changes and summer the least, but all but a handful of older folks have forgotten that winter temperatures of -30°C were common in the 1990s, and -40°C was not uncommon. Now, extreme temperatures of -40°C or lower occur only once every five years or so. January temperatures have increased by 5°C to 8°C throughout the southern prairies. The temperatures of January 2040 are very similar to the Marches of the 1990s. Winter is much easier to survive now. But, of course, people still complain about the cold.

Are milder winters on the way? (Greg Pender)

The snowfall season has shrunk by almost two months, but snowfall amounts have grown nearly 150 percent. The snow cover is deeper and more dependable, giving a boost to winter recreation. Cross-country skis, snowmobiles, and snow twisters are selling like hot cakes. But snow-removal budgets have increased, despite the shorter season.

Fewer people are travelling south to escape the harsh prairie winter—a trend which has depressed the tourist trade in places such as Arizona. The state was already suffering from the searing hot summers and hyperaridity brought on by global warming. Now more and more people are spending their tourist time on the prairies—the "Arizona of the North."

Migrating Plants and Animals

Temperatures have increased the most on the prairies and in the western boreal forest, as climatologists projected in the 1980s. In fact, these areas have experienced the most intense warming in all of Canada. National and provincial parks have been severely affected. Not only have increasing human populations put more pressure on the parks, but the changing climate has forced many parts of the ecosystems northward. The migration was facilitated by well-planned corridors to northern parks, so many plants and animals were saved. We also gained plants, animals, and insects that were moving northward with the climate.

Several tree types, however, did not migrate fast enough and could not find suitable climatic habitats. The Cypress Hills Park was especially hard hit because of its unique altitude. As tree varieties migrated upward along the slopes, they were soon faced with nowhere to go. Several rare and endangered species were squeezed out by climatic changes.

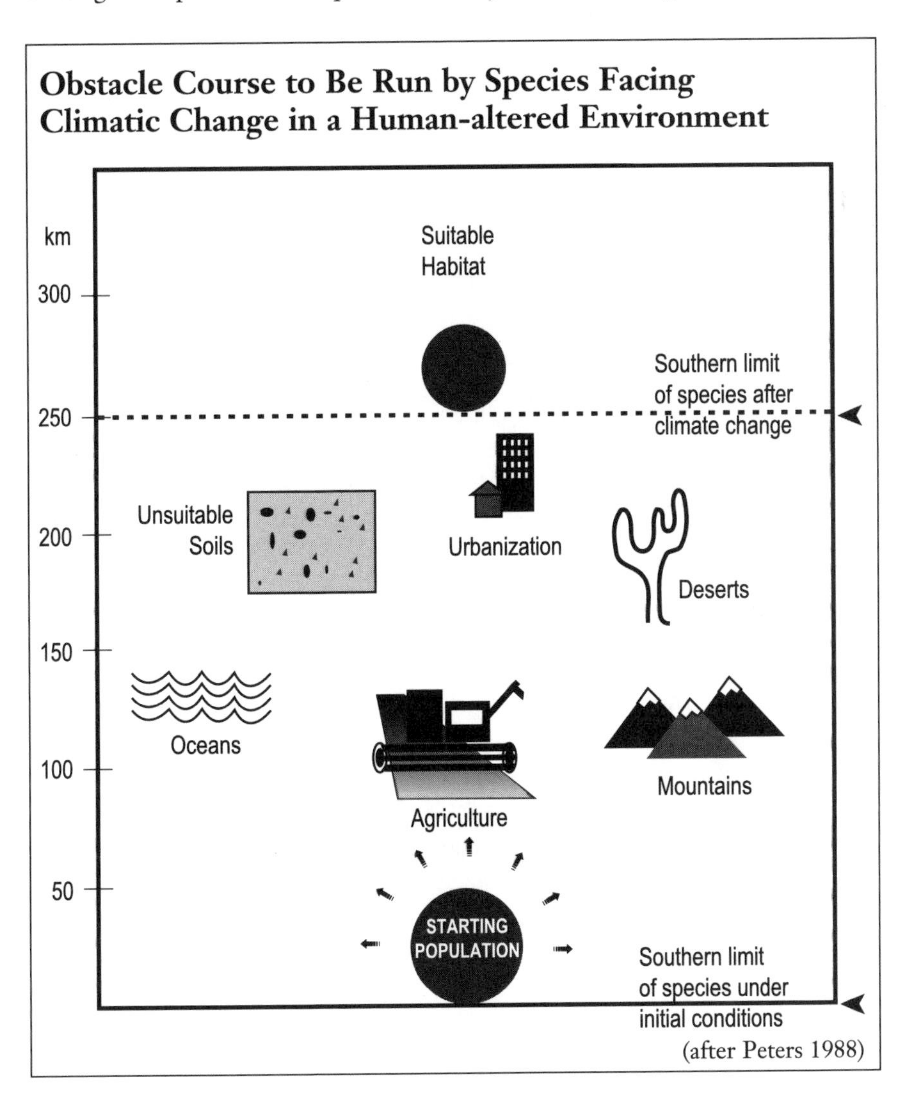

Obstacle Course to Be Run by Species Facing Climatic Change in a Human-altered Environment

(after Peters 1988)

Ecosystems are sensitive to small changes in average temperatures. It only took a 2°C increase in the parkland to convert it to grassland. After the big aspen fire of 2010 in Manitoba, reforestation efforts have been severely challenged by the droughts.

Wanted: Adaptation and Innovation

Many of the adverse effects of the changing climate had been dealt with or were being prepared for by 2040. This adaptation was thanks to the information and expertise provided by the Prairie Climatic Change Adaptation Network. The Network had been set up in Saskatchewan by 1997, after the International Intergovernmental Panel on Climate Change (IPCC) had established that global climatic changes were not due to natural causes alone. Changes related to human activities had been identified.

Prairie researchers had already been working on the topic of climatic change, its impacts, and adaptation strategies for years. By the early 1990s, this region was considered a focus of effort in the climatic impact and adaptation assessment field for several reasons:

- The importance of agriculture to the area, to Canada, and to the world. Prairie agriculture is strongly linked to weather and climate.
- The prairies' sensitivity and vulnerability to drought and other climatic events known to cause severe environmental, economic, and social effects.
- The common occurrence of many climatic hazards in the area.
- The expectation that high-latitude, mid-continental regions such as the prairies were to experience the largest climatic changes under continued global warming.
- The occurrence of the greatest warming trend in Canada in the past century in a broad corridor from the prairie provinces northward into the Mackenzie District.

By the closing years of the twentieth century, it was clear that the prairies were an important natural laboratory in vital need of further climatic impact and adaptation work.

Back to the Present

Predicting the future is risky in any field, and none of the situations described above may happen. But current assessments suggest that they are one of a possible set of future climates and their effects. We must consider what the new climates might be and what they will mean to all of us who live here in the future.

Although it was once thought that climates changed slowly, if at all, mounting evidence indicates that they can change—and have changed—rapidly. Recently, paleoscientists studying evidence from glaciers and lake sediments found that extensive high northern areas have undergone temperature changes of several degrees in a period of years or decades, and that these abrupt, major changes have occurred many times in the past 130,000 years. Drought and flood episodes on the great plains of North America, for example, have been much more severe in the more distant past than any we have experienced in the past 150 years.

Scientists have been able to compile trends of major climatic changes over millions of years. Not surprisingly, the lowest temperatures have occurred during global glaciations at approximately 100,000-year intervals over the past 800 millennia. Interglacial periods with warming of 4°C to 6°C in average temperatures followed the ice ages. These seemingly small global temperature changes had enormous effects in terms of seasonal and daily weather.

Instrument records provide the best source of information, but data only exists for the past century or so. According to these data, average global temperatures have shown a general warming trend since the late 1890s. A brief cool period from about 1940 to the mid-1960s was superimposed on this almost century-long warming trend. But even though temperatures were low for that period, they were still higher than those at the start of the instrumental record in the late 1890s.

Since the 1960s, the earth's average temperature has continued to rise. The increase over the past hundred years globally has been about 0.3°C to 0.6°C. This may seem small, but in global and climatological terms it is large. Many of the highest average yearly temperatures for the earth have clustered in the late 1980s and the 1990s. In terms of global surface temperature, 1995 has been the warmest year so far, with a temperature of 0.4°C higher than the 1961–1990 average[1]. The second warmest year was 1990, which saw an increase of 0.36°C above average. As of 1996 the ten warmest years on record have all occurred since 1980, which inaugurated the hottest decade globally in the past 130 years. The eruption of Mount

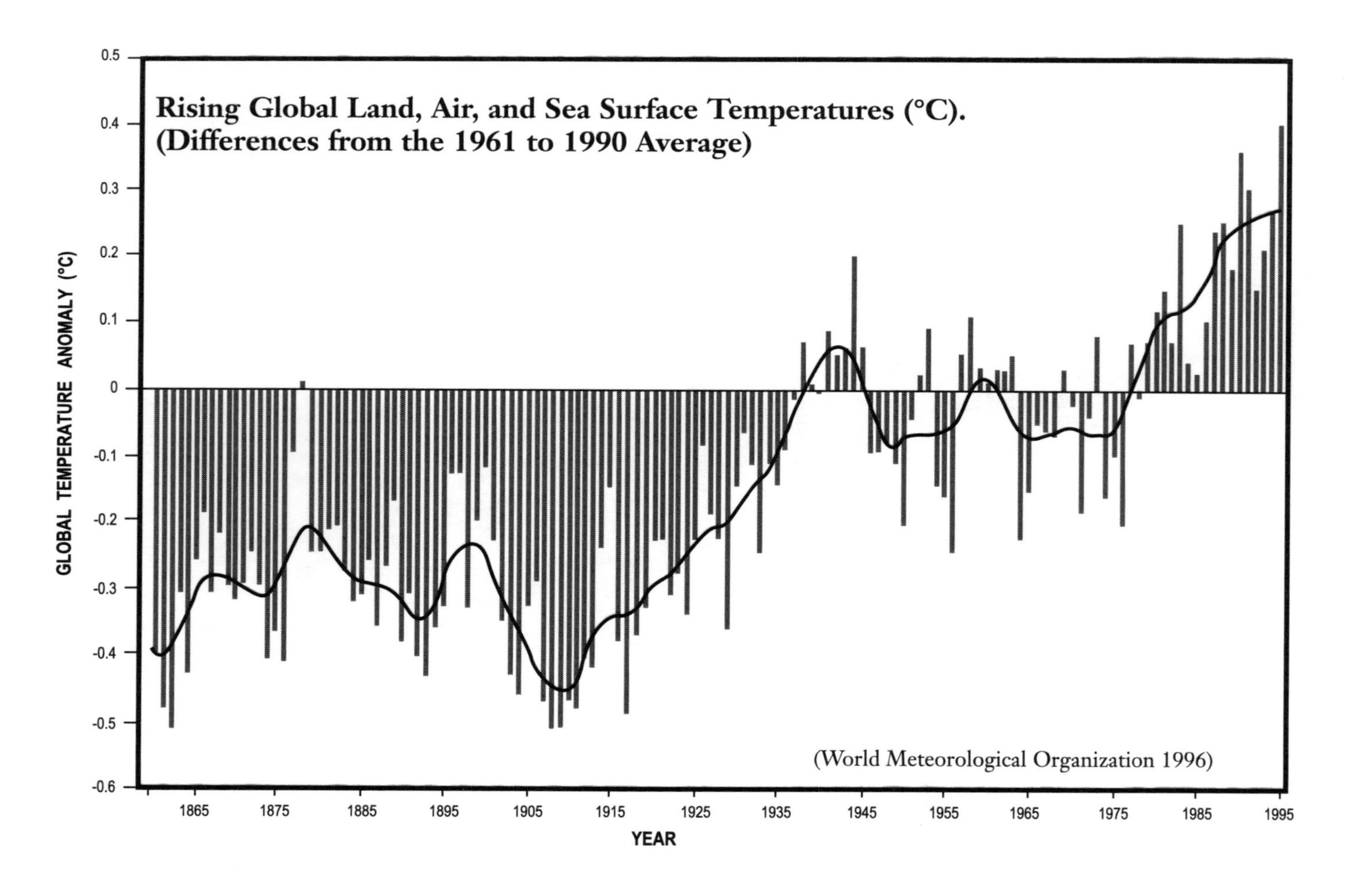
Rising Global Land, Air, and Sea Surface Temperatures (°C).
(Differences from the 1961 to 1990 Average)
GLOBAL TEMPERATURE ANOMALY (°C)
0.5
0.4
0.3
0.2
0.1
0
-0.1
-0.2
-0.3
-0.4
-0.5
-0.6
(World Meteorological Organization 1996)
1865
1875
1885
1895
1905
1915
1925
1935
1945
1955
1965
1975
1985
1995
YEAR

Canada's Annual Temperature (°C), Precipitation (mm), and Cloudiness (percent) Trends, 1953–1992

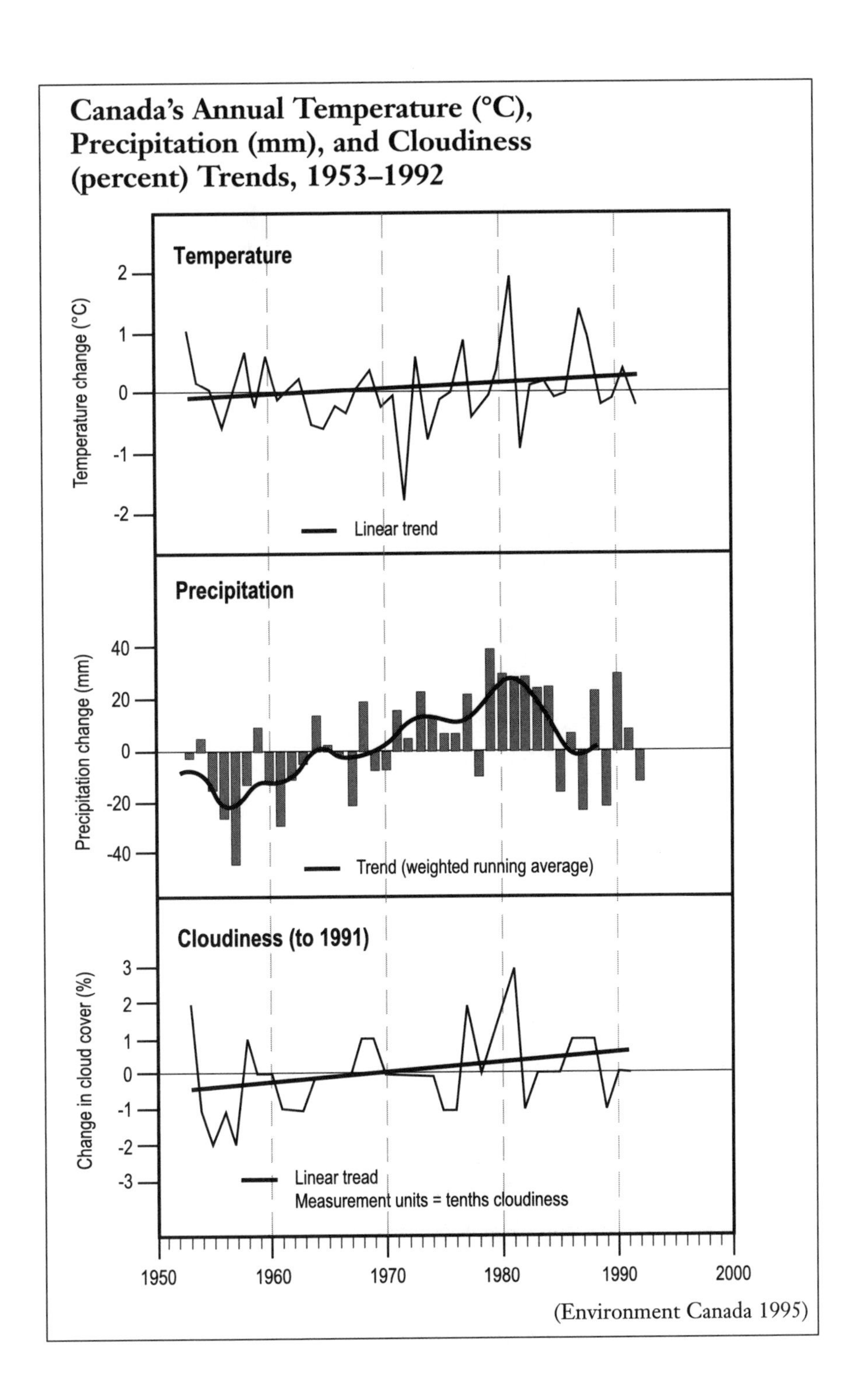

(Environment Canada 1995)

Changing Annual Surface Temperature (°C per decade), 1961–1990

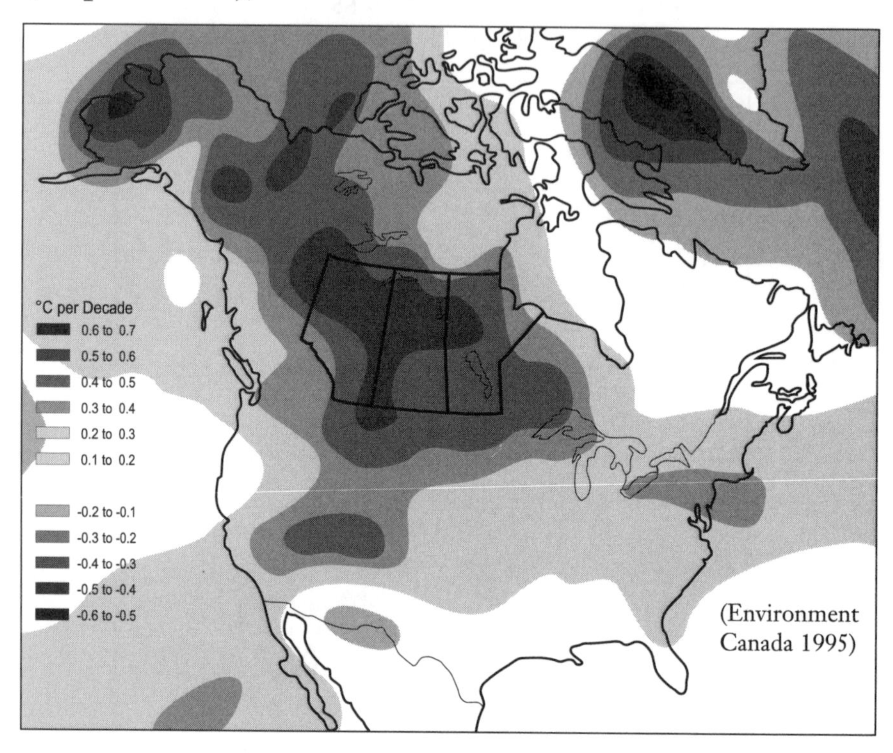

(Environment Canada 1995)

Pinatubo in 1991 temporarily flattened the steep warming trend, but that effect disappeared after about two years.

A small change in global temperature represents a large change in a region. The average global temperature difference between an ice age and an inter-glacial age, for example, is only about 4°C to 6°C. An increase of 1°C globally would result in regional temperatures the earth has not experienced for at least 160,000 years.

Ecosystems are particularly sensitive to small changes in average temperature. When you consider that the average annual temperature of the parklands is only about 2°C higher than that of the forest region of the prairie provinces, you realize how little it would take to change the forest to parkland. Further, a small increase in average monthly temperatures for a certain area can make a huge difference in extreme daily temperatures. An increase of only 4°C to 5°C in mean monthly temperatures in May 1988 resulted both in severe droughts and in temperatures of over 40°C. Many places experienced record-breaking heat waves that year.

El Niño Strikes Again

No weather phenomenon has been accused of more weather crimes than the infamous El Niño. Blaming weather anomalies on El Niño may be convenient (and, at times, deserving), but it is not always correct. Local conditions of drought, flood, plant cover, or snowcover—to name a few—also play a part in prairie weather, as do ocean temperatures in areas such as the North Pacific or Arctic Oceans.

El Niño (short for El Niño-Southern Oscillation, or ENSO) is an oceanic and atmospheric phenomenon that involves unusually warm temperatures in the central to eastern equatorial Pacific Ocean. Centuries ago, people fishing off the coast of Ecuador and Peru around Christmas time noticed unusual ocean current and temperature conditions. Hence the name El Niño, Spanish for "the Christ Child."

About every two to seven years, ENSO settles in and tends to cause anomalous weather, and sometimes havoc, in many areas. The effects are most apparent in the Indo-Pacific Ocean, but climatic impacts occur around the world during strong ENSOs. The worldwide catastrophic weather changes brought by ENSO include flooding in northern coastal Peru (usually a desert), and in California, and droughts in the southwestern United States, western Europe, Sahelian Africa, India, northern China, Australasia, and the Caribbean.

So what does an ENSO mean for the Canadian prairies? It can be something to look forward to, as it usually brings warmer-than-normal winter temperatures. The 1997–98 ENSO was one of the strongest on record, rivalling even the major ENSO of 1982. The 1997 event brought the strangest prairie winter, if you could call it "winter," with fall-like temperatures until December and a brown Christmas.

During the severe 1988 drought, the opposite of El Niño, or La Niña (cold and dry conditions over eastern tropical Pacific), occurred during the summer. According to prairie researchers, prairie droughts may be initiated by certain temperature patterns in the North Pacific Ocean, as well as in the tropical ocean. This drought pattern results from the combination of a huge pool of unusually cold water in the east-central North Pacific Ocean, and a warm pool along the central-west coast of North America. Other factors, beside sea surface temperatures, can influence extreme climatic events; these include regional and local soil moisture and snow-cover characteristics.

(For further reading refer to Allan, R., J. Lindesay, and D. Parker. 1996. *El Niño Southern Oscillation and Climatic Variability*. Collingwood, Australia: CSIRO Publishing.)

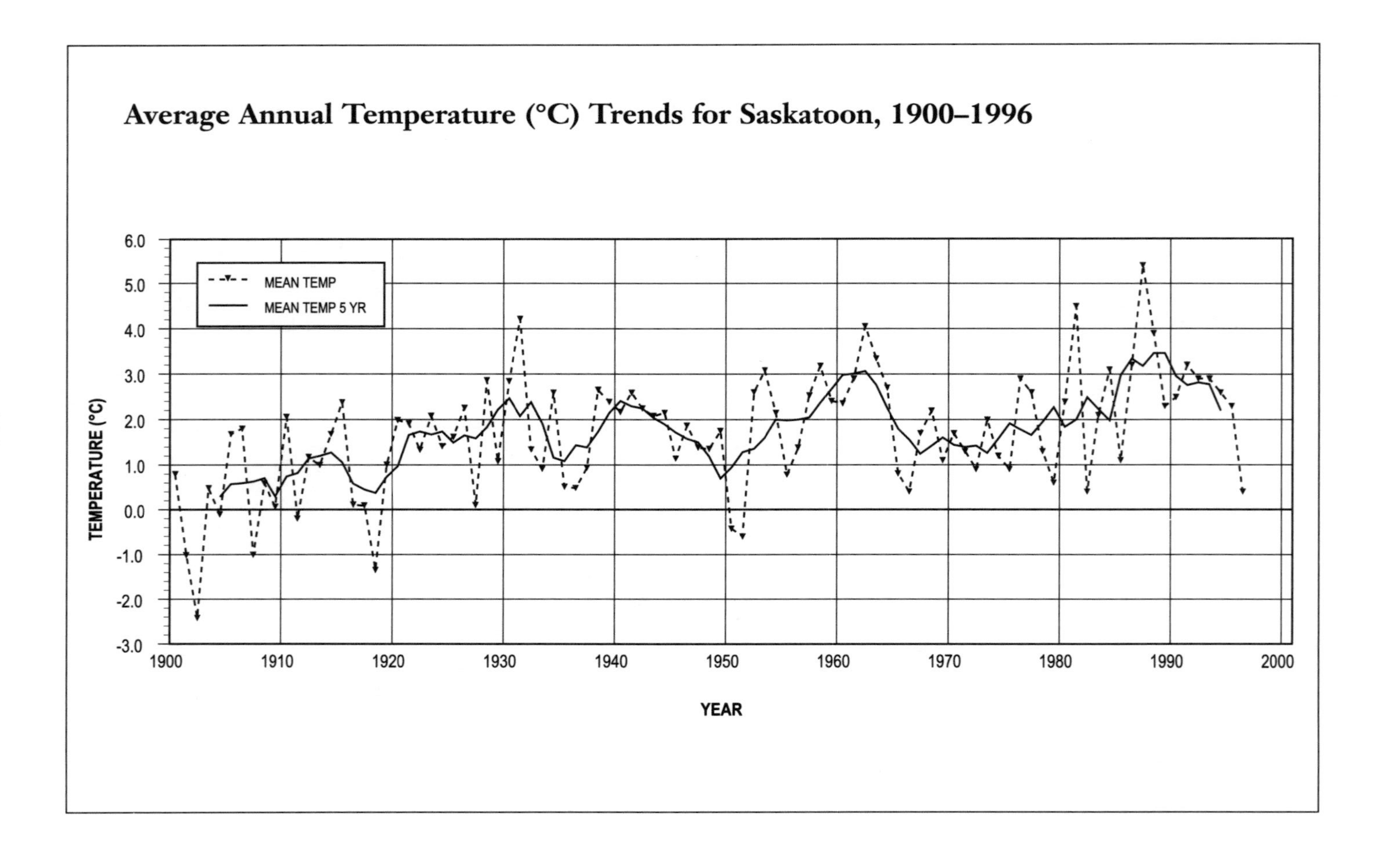
Average Annual Temperature (°C) Trends for Saskatoon, 1900–1996
MEAN TEMP
MEAN TEMP 5 YR
TEMPERATURE (°C)
6.0
5.0
4.0
3.0
2.0
1.0
0.0
-1.0
-2.0
-3.0
1900
1910
1920
1930
1940
1950
1960
1970
1980
1990
2000
YEAR

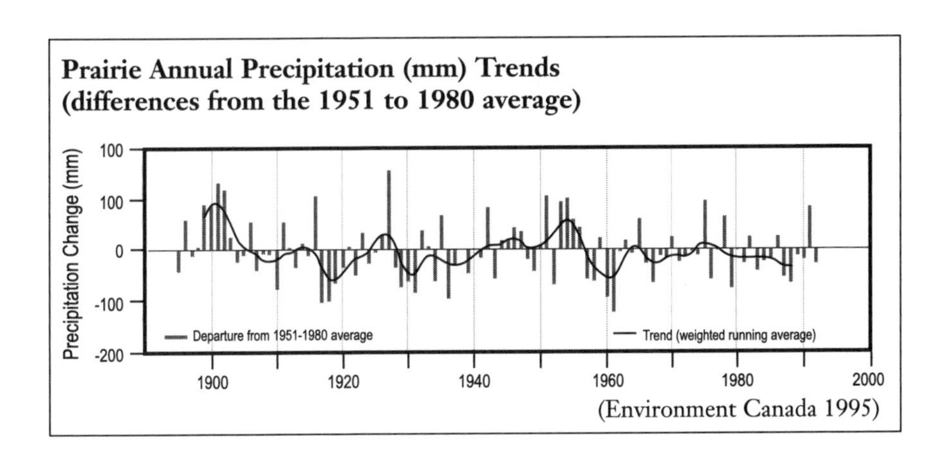

Prairie Annual Precipitation (mm) Trends (differences from the 1951 to 1980 average)

(Environment Canada 1995)

Canada's climate has kept pace with global warming; temperatures have increased by over 1°C in the past hundred-year period. This warming has not been uniform in time, area, or rate. Rather, there have been three distinct phases: 1) a warming from the 1890s to the 1940s; 2) a cooling to stable trend from the 1940s to the early 1970s, with a period of a few years as cold as or colder than the late 1890s; and 3) a resumption of warming from the 1970s to the present.

Interestingly, the greatest rate of warming has occurred at night—a trend that has also been observed in other countries, such as the United States, China, and Australia—and during winter and spring. The seasonal contrast has also been experienced in other countries of the northern hemisphere, although temperature changes have varied considerably from region to region. The greatest warming in Canada over the past hundred years occurred in a broad corridor running northwest over southern Saskatchewan to the Mackenzie District. Baffin Island and Ellesmere Island in the far north experienced a moderate cooling of 0.6°C.

Some of the most dramatic climatic changes in Canada, and perhaps the world, have occurred on the prairies. Individual prairie climate stations reflect this trend over their periods of record. For example, a Saskatoon station that began recording climate data in 1889 shows the prairie warming trend to the 1940s; a decreasing trend to around 1950; an increasing trend to the early 1960s; a decreasing trend to the mid 1970s; and an increase to the present. Underlying these fluctuations is an overall increase in temperature: each successive cooling trend was not as cool as the previous one. The cooling period of 1963 to 1966, for example, was relatively much warmer than the cool period of the early 1900s.

Brandon, on the other hand, doesn't fit the general pattern of the prairie region. Although July temperatures and growing-degree days did increase, the growing season length and January temperatures do not exhibit any noticeable trends. The frost season seems to have shifted slightly toward winter, with later spring frosts and later fall frosts.

Changes in climate, of course, include not just temperature, but precipitation, snow cover, cloudiness, wind, and other elements. These are trickier to examine because they are more difficult to measure. However, trends in a few elements, e.g., precipitation and cloudiness, have been documented.

Canada has been getting cloudier, especially since the 1970s, and it has probably become a wetter nation since 1948. Until about 1964, annual national precipitation was generally below the long-term average. An upward trend dominated the next stage, with especially high precipitation in the early to mid-1980s. This was followed by a variable period. The pattern fits the global trend.

The plains, in contrast, have been getting drier. In fact, the prairie region and the mountains of southern British Columbia are the only regions out of the eleven climatic regions of Canada that show a downward trend in annual precipitation. The annual precipitation pattern on the prairies was high in the early 1900s. Then low periods occurred, especially during the late 1910s, 1928 to about 1940, 1957 to about 1962, and 1979 through much of the 1980s.

The roller-coaster of fluctuations in precipitation on the prairies shows up at several stations: Saskatoon's annual precipitation has ranged from drought years with only 224 millimetres to monsoon-like years with values over 550 millimetres. Brandon's precipitation is "all over the map" as summer values (May to September) ranged from as low as 105 millimetres in 1967 to as high as 524 millimetres in 1935. Prairie precipitation is the most variable in Canada.

Snow cover, too, has changed. Two basic aspects of snow catch people's attention: length of snow-cover season and snow depth. The snow-cover season has been shrinking over the past thirty years or so over a large area from the west coast of Canada across the southern prairies to the Great Lakes.

The southern prairies have among the largest rates of decrease, at more than one day less of snow cover each year. This change was concentrated in the past twenty years and is related to earlier springs. The entire northern hemisphere shares this trend. In contrast, a couple of smaller areas of Canada, such as Newfoundland, have experienced an expanding snow cover season.

Key Greenhouse Gases Affected by Human Activities

	Carbon Dioxide	Methane	Nitrous Oxides	Choro-fluoro-carbons	CFC substitute (hcfc-22)	Perfluoro-carbon (CF_4)
Pre-industrial concentration	280 ppmv*	700 ppbv*	275 ppbv	zero	zero	zero
Concentration in 1992	355 ppmv	1,714 ppbv	311 ppbv	503 pptv*	105 pptv	70 pptv
Recent rate of Concentration change	1.5 ppmv/yr	13 ppbv/yr	0.75 ppbv/yr	18-20 pptv/yr	7-8 pptv/yr	1.1-1.3 pptv/yr
Concentration change per year (over 1980's)	0.4%/yr	0.8%/yr	0.25%/yr	4%/yr	7%/yr	2%/yr
Atmospheric lifetime (years)	(50-200)**	(12-17)***	120	102	13.3	50,000

*1 ppmv = 1 part per million by volume, 1 ppbv = 1 part per billion by volume, and 1 pptv = 1 part per trillion (million million) by volume

**No single lifetime for CO_2 can be defined because of the different rates of uptake by different sink processes.

***This has been defined as an adjustment time which takes into account the indirect effect of methane on its own lifetime.

(Intergovernmental Panel on Climate Change 1996)

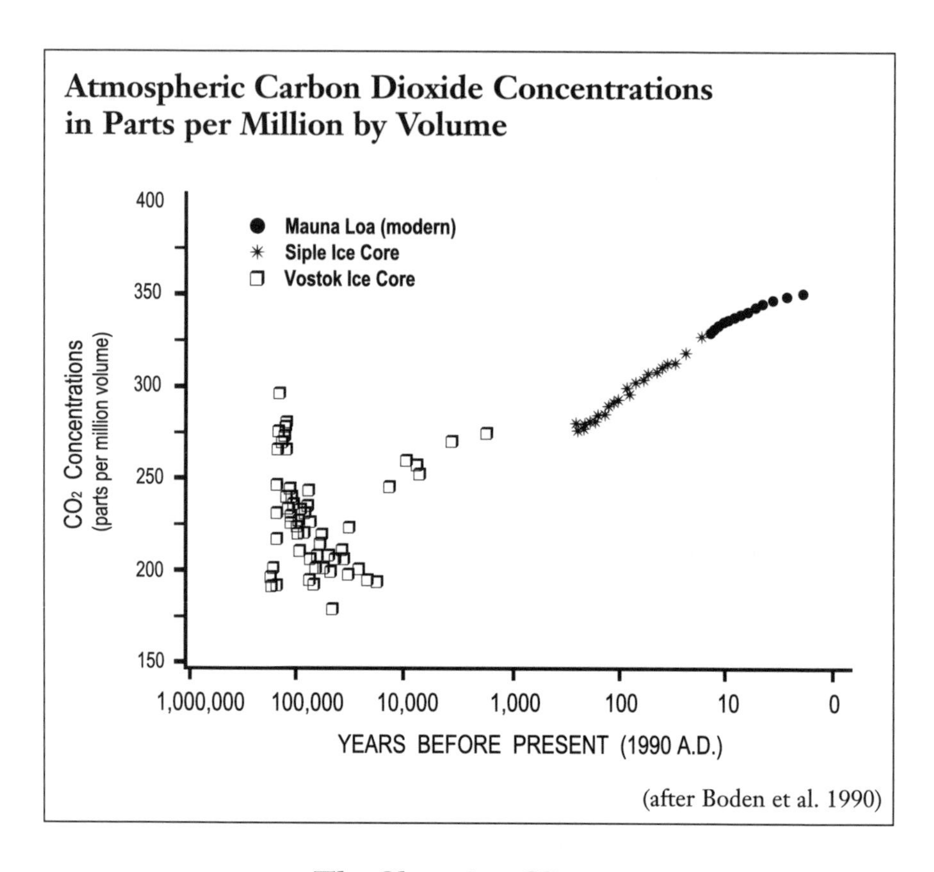

The Changing Climate

Many factors act in complex ways both to produce and to alter the earth's climates. These include changes in:

- solar energy related to changes in the sun's radiation and the earth's orbit;
- atmospheric particles related to pollution and volcanic activity, among others;
- reflectivity of the earth's surface (how bright it appears from space);
- trace gases in the atmosphere (the greenhouse effect).

Some of these factors have a cooling effect; others have a warming effect; some can have either. The time scales are also vastly different. Orbital changes could affect climate over tens of thousands of years, whereas the effect of dust particles in the atmosphere could last from a few days to a few years.

Human activities, too, have changed and are changing the earth's climates. The burning of fossil fuels, deforestation, and agricultural practices,

Budgeting for Heat

The earth's heat budget is like a system of deposits and withdrawals. Radiation comes in from the sun. About half of it is absorbed by the earth's surface. Of the other half, approximately thirty-one percent is reflected back to space, while nineteen percent is absorbed by ozone and fine particles suspended in the atmosphere.

The earth's surface and lower atmosphere are heated by the sun's rays. They, in turn, give off this energy as heat. Clouds and greenhouse gases are the two major obstacles blocking this outgoing energy. The fact that the atmosphere is "powered" this way by the earth's surface rather than directly from the sun is fundamental to understanding weather and climate.

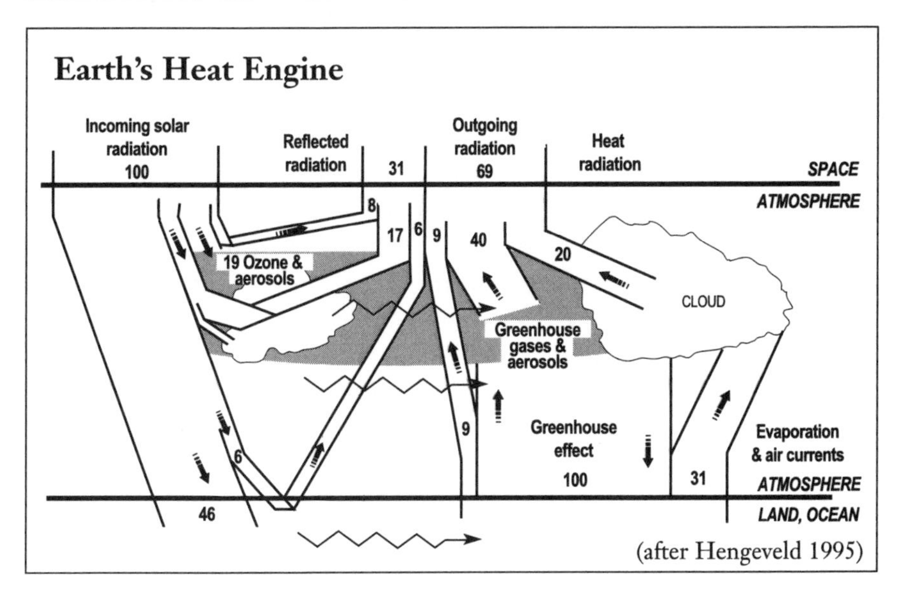

among many other human activities, add greenhouse gases to the atmosphere, including carbon dioxide, methane, nitrous oxide, and chlorofluorocarbons. The chemical composition of the atmosphere is quite different now than it was just a century and a half ago. These trace gases, although they occur in relatively small amounts, are changing the atmosphere at an unprecedented rate.

Six billion tonnes of carbon are added to the atmosphere annually from human activities—an increase of eleven percent in thirty years. The current level of carbon dioxide is greater than the highest values of the past 160,000 years by more than twenty percent. Sources of carbon dioxide emissions include the burning of fossil fuel and of forests.

Other greenhouse gases are also increasing. Methane has doubled over its pre-industrial levels. Sources of methane emissions include landfills and rice paddies. Nitrous oxide amounts have gone up by five to ten percent. Sources of nitrous oxide include agricultural fertilizers and synthetic chemicals such as chlorofluorocarbons (CFCs) that are not naturally present in the atmosphere. CFCs are used in refrigerators, air conditioners, and as propellants for aerosol spray cans.

These atmospheric changes have been well documented for years. Monitoring stations at Mauna Loa, Hawaii; Alert, North West Territories; Sable Island, Nova Scotia; and other sites around the world regularly measure the concentrations of greenhouse gases in the atmosphere. The information is compiled and analyzed by organizations such as Environment Canada and the United States Carbon Dioxide Information Analysis Centre.

The Greenhouse Effect—Goldilocks' Earth

The term "greenhouse effect," one of the first concepts taught to climate students, has sprung out of the textbooks into everyday usage. It is a term used to describe the roles of water vapour, carbon dioxide, methane, and other trace gases in the earth's heat budget. These gases let sunlight through the earth's atmosphere, but trap a large amount of heat that tries to escape to space. They are referred to as "greenhouse gases" because their effect on the earth is similar to the heating processes of a greenhouse.

The windows of a greenhouse let in sunlight, which warms up the plants and pots and soil inside. These objects then give off heat. The glass in the roof and sides of the greenhouse is relatively transparent to solar energy, but relatively opaque to re-radiated heat energy. Heat is thus trapped in the greenhouse. The same thing occurs in a parked car on a sunny day.

Greenhouse gases similarly allow sunlight through the earth's atmosphere. They are relatively transparent to incoming solar energy, which arrives as shortwave, visible radiation. The sun's energy heats the earth's surface, which in turn re-radiates the energy in long-wave form. This long-wave radiation, which would otherwise escape into space, is absorbed by greenhouse gases and heats the atmosphere.

The greenhouse effect keeps the earth about 33°C warmer than it would be otherwise. Without this protective blanket of water vapour, carbon dioxide, and trace chemicals, the planet would have an average surface temper-

Goldilocks' Earth: Mean Planetary Temperatures (°C) for Earth, Mars, and Venus.

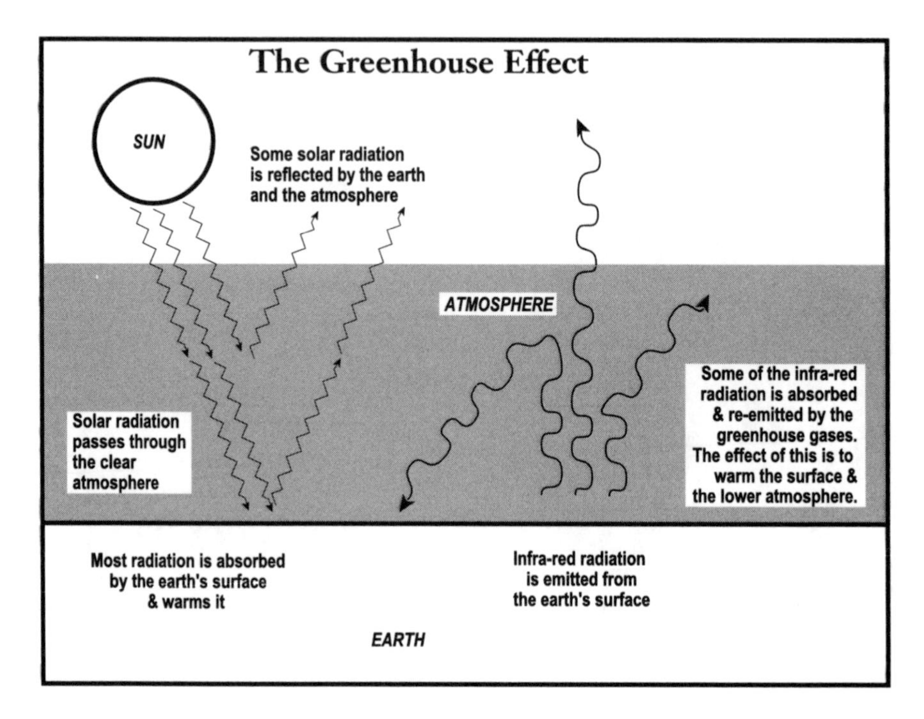

ature of -18°C. Thanks to the natural greenhouse effect, the earth's temperature averages about 15°C.

Unfortunately, it is possible to have too much of a good thing, and too much greenhouse gas leads to global warming. Venus is a good example of a runaway greenhouse effect, while Mars is the opposite. The differences in the atmospheres of these planets make Mars a deep freeze, Venus an oven, and Earth just right.

Global Warming

The term "global warming" is often used to mean warming related to the greenhouse effect. However, global warming has other natural causes and has occurred several times over a long history without human intervention. The most recent assessment from the prestigious Intergovernmental Panel on Climate Change (IPCC 1996), however, is that the climatic changes of the past century were unlikely due entirely to natural causes; they are linked to the global increase of greenhouse gases. The impact of human beings is there in the climatic record. Unprecedented global warming is to be expected in the future.

Temperature changes, changing precipitation patterns, and extreme events could cause disruption in ecosystems and human activities,

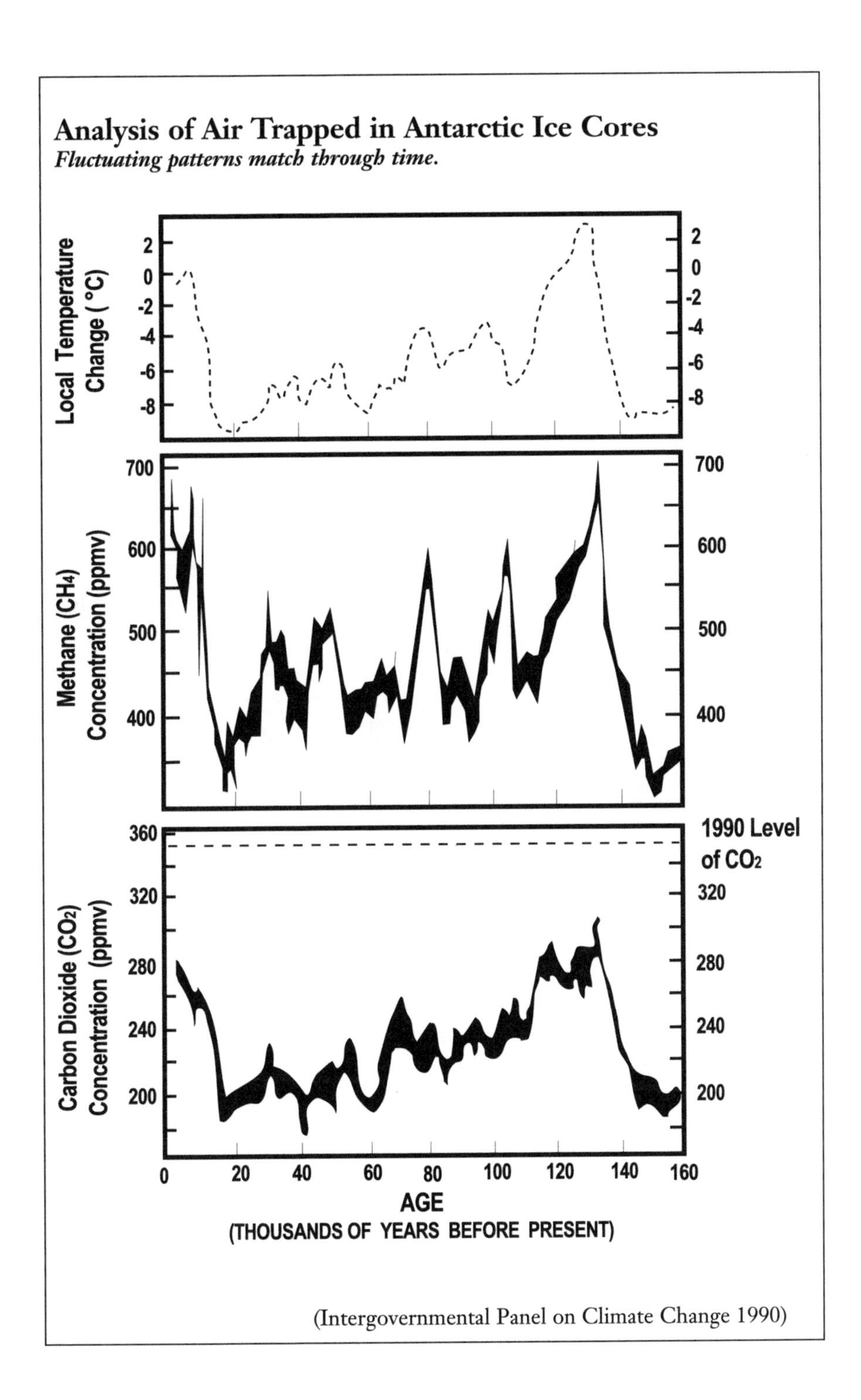

(Intergovernmental Panel on Climate Change 1990)

When Is Global Warming Arriving?

n angry reporter called me on a particularly nasty day during a long cold spell in February, 1995. "Why are we having such a rotten winter?" he demanded. "When is this global warming supposed to arrive?" Perhaps he had been expecting banana trees to spring up in Winnipeg. Did I dare tell him that we have already had some global warming, that earlier winters had been much worse than this one?

I told him that these past cold days, even the winter itself, were weather events, not climate, and they certainly did not indicate climatic change. Climate is a longer-term pattern; weather events are often just noise in the system. Single periods of cold days and hot days are not evidence of global cooling or global warming, even in an area as large as the Canadian prairies. Only shifts in many weather events over thirty or more years could be taken as evidence of change. And for global conditions, it is important to use data from stations that represent the earth, including the oceans, not just one region.

Perhaps I should never have answered the phone. But the reporter did cheer up (a bit). He also wrote an interesting article.

requiring radical adaptation if we are to avoid catastrophic losses.

A significant amount of global warming has occurred in the past 100 years, a trend which has accelerated in the past two decades. But climates are dynamic. They're bound to change naturally, as they have in the past. So what evidence is there that greenhouse gases are responsible? Where's the smoking gun?

Some of the best evidence lies in surprising places. Gas bubbles trapped in glacial ice, for instance, contain carbon dioxide and methane found in the atmosphere of many thousands of years ago. The amounts of these gases closely match the pattern of average temperatures. When greenhouse gases were abundant, temperatures were high, and vice versa.

Mathematical experiments also play an important role in building evidence. General Circulation Models, or GCMs, are like theoretical models of the earth that we can pollute at will and measure how the climate would change. These models are based on fundamental laws of physics and conservation of mass, momentum, and energy.

Researchers have "asked" GCMs what the global and large-area regional climates would look like at the time of increased greenhouse gas. The results of the models agree on many findings:

- increased warming in higher latitudes, especially in late fall and winter;
- increased precipitation in higher latitudes and the tropics throughout the year, and in mid-latitudes in winter;
- increased drying of the earth's surface across large areas of the northern mid-latitudes during the northern summer.

The Intergovernmental Panel on Climate Change (IPCC 1996) reports that a doubling of the earth's greenhouse gases could lead to an increase of more than 1°C to 3°C in the next fifty years—a rate of warming higher than any in the past 10,000 years.

Looking into the future is tricky at the best of times, but our best evidence points to the following changes:

- greatest temperature increases occurring in the winter and spring, with the most changes in the south centring on Saskatchewan;
- increased night-time temperatures;
- smaller, but significant temperature hikes in the summer;
- largest precipitation increases occurring during the winter and spring;
- expected rainfall decreases in the summer in the southern prairies;
- increased growing season and decreased snow-cover seasons and areas;
- greater intensity and frequency of droughts as well as floods.

Evidence is also accumulating in trends related to temperature, precipitation, snow cover, permafrost, and other elements. These are the "smoking guns" we would expect to find in a climate increasingly dominated by the greenhouse effect. Other evidence suspected to be related to global warming includes:

- four El Niños have occurred in the past five years—an unprecedented frequency;
- in the past ten years, the southeast United States have experienced the worst drought in 300 years, while the midwest had its worst flood in recorded history;
- since 1980, Canada has suffered five of the worst forest-fire years in history;
- permafrost has been thawing more rapidly and extensively;
- insurance claims because of natural disasters in the 1980s were ten times higher than in the 1960s, and even higher in 1991 and 1992. Storms were

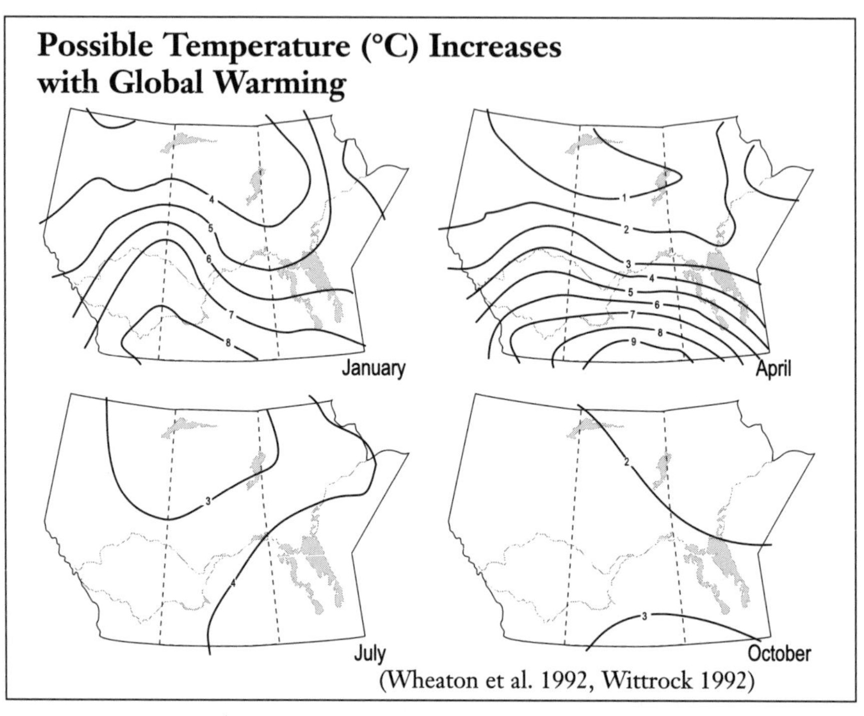
Possible Temperature (°C) Increases with Global Warming
January
April
July
October
(Wheaton et al. 1992, Wittrock 1992)

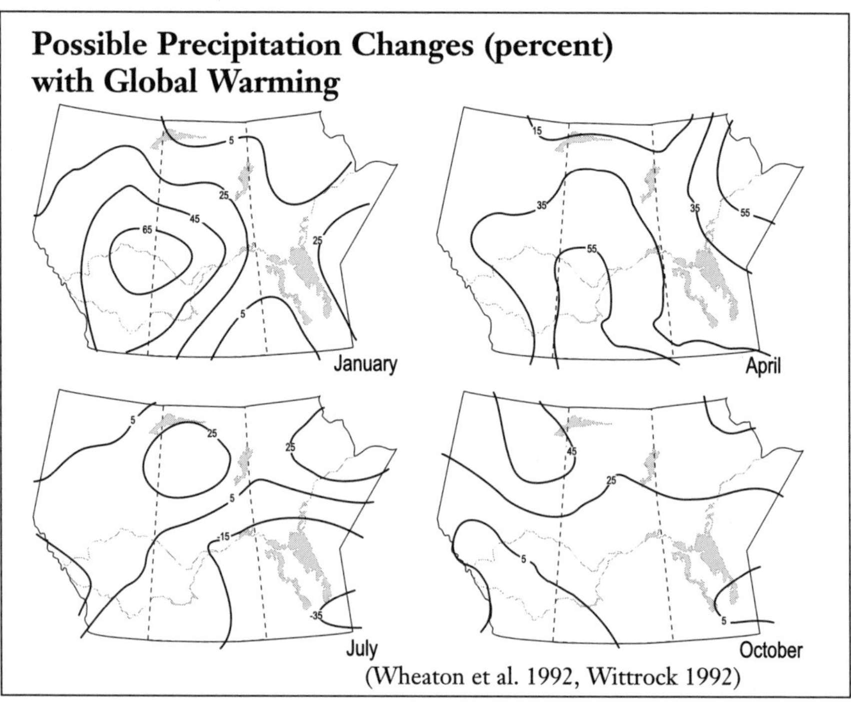
Possible Precipitation Changes (percent) with Global Warming
January
April
July
October
(Wheaton et al. 1992, Wittrock 1992)

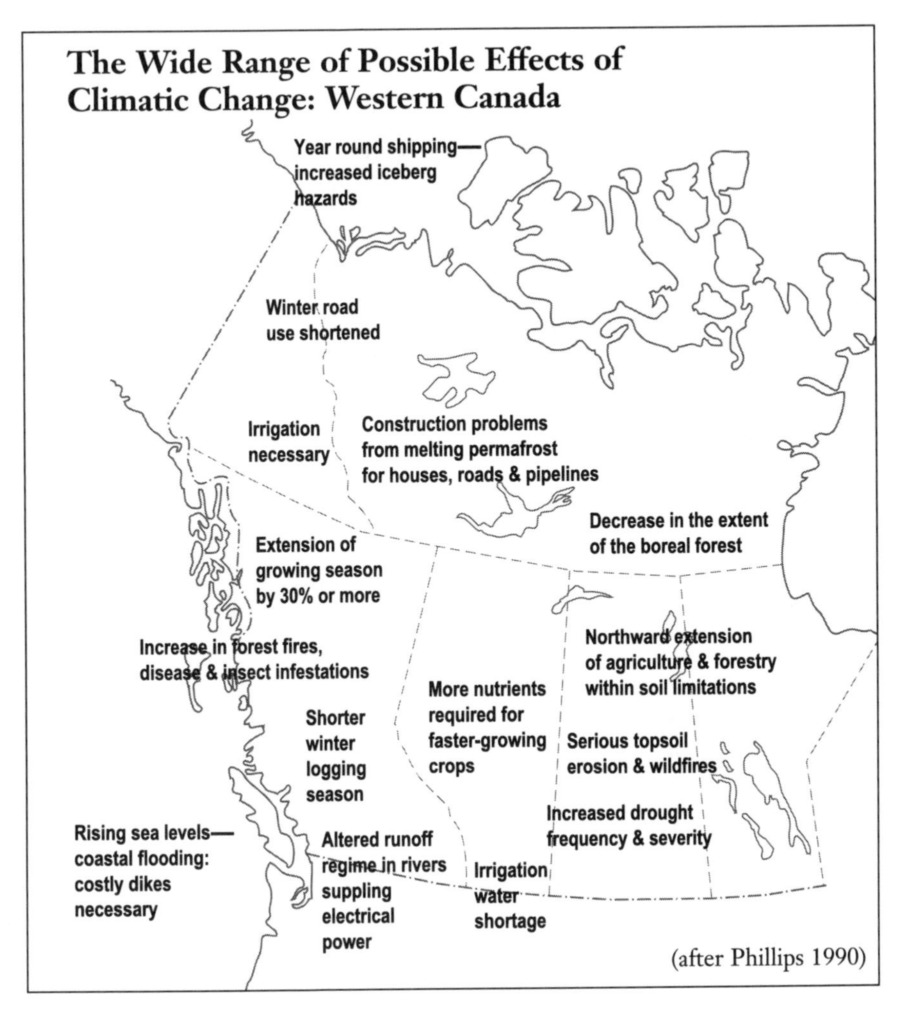

responsible for eighty-eight percent of the natural disasters in the past decade;

- catastrophic windstorms worldwide are increasing year after year;
- northern hemisphere snow-cover has shrunk in the last twenty years;
- a general global deglaciation trend has occurred over the past century.

The potential effects of global warming have not yet been the subject of much research, but we do know that climate affects many aspects of both the natural and the human environments. Part of the problem, of course, is that we have traditionally separated the "human" and the "natural" environments—as if humans were somehow unnatural, or not part of the environment. It is this very separation, in attitude if not in fact, that has allowed us to plunder nature without regard for future consequences. All the research in the world will go for nought if we cannot overcome this attitude. For agriculture, forestry, energy, recreation, tourism, health, transportation, utilities, wetlands, water resources, economics, not to mention human society as a whole, are all affected by the climate in which they function. Because change is expected to be relatively large and rapid, it is reasonable to conclude that the consequences will be major, and some will be disruptive and severe. With good preparation and response, however, other effects will be positive. Only one thing is certain: we should not be surprised by surprises.

As for the prairies, the jury is still out on whether we're in good shape with respect to climate change, or if we're going to run into some tough situations down the road. Much depends on how well we can use the information and how well we are prepared. There are many good points about prairie weather: sunny days, good air quality, fewer ice storms than in Eastern Canada, fewer hot and humid days. It will be interesting and useful to know if and how these conditions might change.

Any speculation about the future is hazardous, but projections are required for adequate planning and preparation. Here are a few of the possible impacts of climatic change in the Canadian prairie provinces:

Air Quality

- increased incidence of smog;
- increased methane emissions;
- changed acid precipitation patterns;
- increased dust from wind erosion;
- increased smoke from forest and grassland fires.

Water Resources

- altered river flows;
- more frequent and severe droughts;
- increased competition for water supplies;
- lower levels in reservoirs.

Agriculture

- longer growing season;
- more heat waves in summer and fewer cold spells in winter;

- changed crop production practices—more drought-tolerant crops planted;
- decreased spring wheat yields, especially with increased droughts, but this could be offset by adaptation and diversification;
- crop yield effects vary with area, management, adaptive responses, etc;
- northward shifts in agriculture where soils are suitable;
- livestock production could be a key to successful adaptation to a changing climate, although sufficient good quality water would be a critical constraint;
- increased air and water quality effects on agricultural ecosystems;
- increased risk of wind erosion;
- increased demand for irrigation.

Agricultural Economics
- Manitoba could experience net positive economic benefits while Saskatchewan and Alberta could experience net negative effects;
- the ability of each province to adapt and thrive is partly related to the range of products produced and their relative prices; this is true of any situation at any time, but even more so at times of crisis;
- land values will increase in northern areas with suitable soils;
- adverse effects may be expected without the adaptation of technology and management;
- insurance premiums would increase.

Forests
- growth and productivity could increase in central and northern regions, especially on favourable sites, and decrease in the south;
- climates will shift northward, putting stress on the forest;
- risks of fire, insects, and diseases could increase, especially with already stressed ecosystems;
- demand for new tree species could increase;
- net loss in the total area of the Western Canadian boreal forest is expected;
- changes in land use (e.g., more agricultural and recreational pressures) will affect multiple use of forest lands.

Wetlands
- increased frequency and duration of low water levels, and decreased size and numbers of wetlands;

- northward expansion may offset southern losses, but changes in wetland types are possible;
- increased human interferences with wetlands.

Tourism and Recreation

- climatic potential for tourism and outdoor recreation will shift northward;
- increases with a longer and warmer summer, depending on adverse effects of hot spells and availability of water resources;
- decreased winter recreation would occur with decreased snow cover season and area;
- indirect effects through climatic impacts on other sectors, including water, agriculture, forestry, wildlife, and the economy.

Energy

- changed renewable sources of energy;
- increased warm season air conditioning demand and decreased cold season heating demand;
- cooling problems for thermal power plants;
- changed hydropower production related to changes in river flows;
- increased costs of energy production.

Human Health

- increased heat stress in summer, but decreased cold stress in winter;
- northward shift in ranges of vector-borne infectious diseases;

Why Do Scientists Focus on the Negative?

To answer this question, we first need to ask if it's scientists who focus on the negative, or the media who report the scientists' findings. Scientists will usually try to report all their findings. Experimental results are often surprising, and sometimes their outcomes, positive or negative, cannot be easily predicted. As for planning for the future, it is often more useful to look for the negatives—the potholes in the road—in order to avoid them. It is much more difficult, for example, to deal with an unexpected expense than with a surprise lottery winning.

- increased air and water quality effects;
- indirect effects through agriculture, recreation, occupational changes. For example, a longer warm season will permit a longer outdoor recreational season.

But What Can We Do?

If we are in tune with the climate, we will be able both to reduce the adverse effects of climatic change and to benefit from its positive effects. To do that, we must improve our level of knowledge and preparedness. We must learn about the climate: how it affects our activities, how our activities affect it, and how best to adapt. Adaptability determines a region's or a nation's vulnerability. Adaptability, in turn, is affected by the adequacy of resources—diversity in both plant and animal species is critical—and by national, provincial, and local policies, which can act either as barriers to or as incentives for knowledge and change.

The guide to Canadian action on global warming is called the "National Action Strategy on Global Warming," which was developed to guide "federal, provincial and municipal governments, all sectors of the Canadian economy and all Canadians in identifying their own actions to work towards national goals" (Government of Canada, 1995). The three basic goals of the plan are:

- limiting and reducing the emissions of greenhouse gases to 1990 levels by the year 2000;
- anticipating and preparing for potential climatic changes which Canada may experience as a result of global warming; and
- improving scientific understanding and predictive capabilities with respect to climatic change.

The following actions address one or more of these goals.

Actions for the Public

- Use public transportation whenever possible. When it's not, drive more fuel-efficient or cleaner-fuel (propane, natural gas) vehicles. Every effort at energy conservation will result in reduced emissions of greenhouse gases into the atmosphere.
- Conserve electricity; use it more efficiently.
- Use alternative energy sources—person power (e.g., bicycles), solar panels, wind energy, geothermal energy.
- Do not use products that contain chlorofluorocarbons (CFCs) or that use CFCs in their production. CFCs are not only powerful greenhouse

gases, but they are implicated in the depletion of the ozone layer.

- Use the four Rs: Reduce, Reuse, Recycle, and Recover. Reducing garbage will help reduce a serious greenhouse gas emission problem. The rising piles of garbage produced by growing populations contribute to the expansion of landfills, which are a major source of methane.
- Conserve soil; soil erosion leads to loss of carbon from the soil.
- Plant trees; preserve forests and grasslands.
- Keep informed; improve your knowledge of this and related issues.
- Become an environmental scientist to help improve the knowledge base.
- Become an environmentally aware decision- or policy-maker.
- Inform politicians of your concerns and let them know what you want done about them.
- Find out more about climatic change and variability and their effects.
- Improve our preparations for and adaptations to climatic change and variability.

Actions for Scientists and Engineers

- Support policies to reduce net emissions of greenhouse gases.
- Monitor the environment (atmosphere, land, vegetation, water—all sectors of the ecosystem) and evaluate trends.
- Add to and improve the sum of human knowledge of the effects of climate on the environment and on resource sectors, and of adaptation strategies.
- Contribute to and improve global understanding of the environment and of the way humans interact with it.
- Improve the public capability to prepare for and adapt to climatic change.
- Communicate with the public.
- Develop alternatives to CFCs; develop alternative energy sources and improve conservation methods.
- Improve electricity demand management.

Actions for Government and Non-Government Agencies

- Stimulate research on climatic change, impact, and adaptation.
- Support environmental conservation measures.

Actions for Everyone

- Increase the amount of energy derived from sources that emit less carbon dioxide. Natural gas, for example, emits less CO_2 than coal. Non-fossil energy sources include hydro, geothermal, wind, solar, and nuclear power.

- Make motor vehicles more energy-efficient. Increase the use of public transit and car pools.
- Use electronic communication—move information, not people.
- Establish higher standards for the efficiency of major electrical appliances.
- Minimize energy waste caused by inefficient heating, lighting, and air-conditioning systems.
- Reduce the destruction of forests and the consequent burning of timber and brush.
- Encourage reforestation and afforestation.
- Spread the use of advanced processes for the production of basic materials and the disposal of waste.
- Prohibit the use of CFCs.
- Increase research on ways to minimize greenhouse gas effects.
- Maximize public awareness of the problem.
- Phase out synthetic gases that escalate the greenhouse effect.
- Adopt all energy conservation measures that are economically attractive.
- Provide economic and practical incentives for individuals to take into account the adverse impacts of their decisions and lifestyles on the environment. For instance, if provincial governments included engine size and fuel consumption among the factors determining annual licensing fees, drivers of smaller, more fuel efficient cars would be rewarded while drivers of gas guzzlers would be penalized.
- Work with developing countries to adopt energy-efficient technologies.
- Provide compensation, where appropriate, to those who suffer harm as a result of human-induced global warming.

Although these lists cover many of the major actions regarding this issue, they are by no means exhaustive. Many more are available, and many more are being developed and implemented.

Although the climate change challenge seems forbidding, there are many things we can do not only to reduce the amount of change, but to adapt to it. The Swift Current innovators mentioned earlier are just one example to show us the way.

References

Boden, T.G., P. Kanciruk, and M.P. Farrell. 1990. *Trends +90—A compendium of data on global change*. Oak Ridge, TN: Carbon Dioxide Information Analysis Center.

Bootsma, A. 1994. Long term (100 year) trends for agriculture at selected locations in Canada. *Climate Change* 26:65–88.

Environment Canada, Atmospheric Environment Service. 1995. *The state of Canada's climate: Monitoring variability and change*. State of the Environment Report No. 95-1. Downsview, Ontario: Environment Canada, Atmospheric Environment Service.

Government of Canada. 1995. *Canada's national action program on climate change—1995*. Ottawa: Government of Canada.

Hengeveld, H. 1995. *Understanding atmospheric change: A survey of the background science and implications of climate change and ozone depletion*. 2d ed. Downsview, ON: Environment Canada, Atmospheric Environment Service.

Intergovernmental Panel on Climate Change (IPCC). 1990. *Climate change: The IPCC scientific assessment*, edited by J. T. Houghton, G.J. Jenkins, and J.J. Ephraums. Cambridge, U.K.: University Press.

______. 1996. *Climate change 1995: The science of climate change*, edited by J.T. Houghton, L.G. Meira Filho, B. A. Callander, N. Harris, A. Kettenberg, and K. Maskell. Cambridge, U.K.: University Press.

Peters, R.L. 1988. Effects of global warming on biological diversity: An overview. In *Proceedings of the first North American conference on preparing for climate change: A cooperative approach*, 169–85. Washington, DC: Climate Institute.

Phillips, D. 1990. *The climate of Canada*. Ottawa: Supply and Services Canada.

______. 1995. News Release. *Summer '95—One for the record*. D. Phillips is Senior Climatologist, Environment Canada, Downsview, ON.

Wheaton, E., J. Konecsni, J. Duerksen, D. Adams, and P. Greenidge, eds. 1993. *Bringing science to life: Global warming, greenhouse effect and the Canadian prairies*. A Science Resource Manual. Saskatchewan Research Council (SRC) Pub. E2900-11E93. Saskatoon: Saskatchewan Research Council.

Wheaton, E.E., V. Wittrock, and G.D.V. Williams, eds. 1992. *Saskatchewan in a warmer world: Preparing for the future*. Saskatchewan Research Council (SRC) Pub. No. E290017E92. Saskatoon: Saskatchewan Research Council.

Williams, G. D. V., R.A. Fautley, K.H. Jones, R.B. Stewart, and E.E. Wheaton, 1988. Estimating effects of climatic change on agriculture in Saskatchewan, Canada. In *The impact of climatic variations on agriculture, Vol. 1, Assessments in cool temperate and cold regions*, edited by M. L. Parry, T. R. Carter, and N. J. Konijn. Boston: Kluwer Academic Publishers.

Wittrock., V. 1992. *Some concerns regarding global warming and the Canadian grassland region*. Saskatoon, SK: Saskatchewan Research Council.

World Meteorological Organization (WMO). 1996. *WMO Statement on the status of the global climate in 1995*. Geneva, Switzerland: World Meteorological Association.

Notes

[1]As of February 1998, the year 1997 was declared the warmest year on record.

Chapter Five

Life as a Climatologist

Most scientists . . . are woefully inept at promoting the value of their work.

Morley Thomas, 1985

Being a climatologist is one of the more unusual jobs in Canada. Most people confuse us with meteorologists, and expect us to be able to forecast the weather. Then they usually complain about it. For my own part, I tell people immediately that I don't forecast the weather, and that I'm glad not to be a meteorologist because they get a lot of flak. A climatologist, I explain, studies the climate and its relationships with the environment, with people, and with their activities, whereas a meteorologist studies the weather and tries to forecast it.

Then I feel compelled to discuss the difference between climate and weather. My short, biased explanation is that the weather is like a daily newspaper, while climate is like a history book—or that weather is like a brick in a building, while climate is the building. Going beyond analogies, weather is the physical state of the atmosphere at a given place and time. We are making a weather statement when we say it is 30°C, sunny, dry, and windy at Killarney, Manitoba, this afternoon, or the forecast for Grey Cup Day is freezing rain and a dangerous wind-chill.

Climate, on the other hand, is a description of the prevailing pattern of weather conditions over a certain area, based on statistics gathered over a long period. These statistics include averages, extremes, and probabilities. You are making a climate statement when you say that the mean annual temperature of Saskatoon is 2°C, with extremes of 41°C and -50°C. You are referring to major climatic regions of the earth when you speak of the tropics, the deserts, the boreal forest, and the tundra. The southwestern prairie is a semi-arid grassland, whereas much of the north is a dry, subhumid forest; each is a climatic region. Climate is the major force affecting the distribution of natural vegetation.

There are not many climatologists in Canada, and only a few in the prairie provinces. But we are not an endangered species. In fact, interest in climatology has increased dramatically with the public's natural interest in climatic change and the greenhouse effect. The main professional society in Canada for meteorologists, climatologists, and oceanographers is the Canadian Meteorological and Oceanographic Society. I belong to the Saskatchewan chapter, which has about twenty members, including students. Only four of us are climatologists. The rest are meteorologists. Not surprisingly, there are no oceanographers in the Saskatchewan chapter.

Who Hires Climatologists?

Most climatologists work for the federal government, or in the geography departments of universities. Climatologists and meteorologists also work for insurance agencies (especially crop insurance), for provincial government departments, and for research organizations such as the Alberta Research Council. Some operate their own consulting businesses.

People use climatic information to ensure that roads, dams, and buildings are designed safely, and that natural resources are managed properly. A great many human activities, both recreational and professional, include a consideration of weather and climate. Climatologists are involved in an ever-increasing range of work. Air quality, agriculture, hydrology, resource management, energy, intelligence (or so I've heard), economics, remote sensing, and various other issues and applications all benefit from the applied research of climatology. The uses of climate information are endless.

Researchers across Canada are interested in the prairie climate. Recent projects at the Saskatchewan Research Council (SRC), some of which are ongoing and some of which are completed, include:

Drought as a Natural Disaster (1995) was a project that defined the disastrous nature of drought and described the history of drought in Canada. We discussed drought monitoring and warning, as well as research needs. The ultimate goal of such work is to reduce the damage drought can cause and develop avoidance strategies. I collaborated with six colleagues from several institutions to prepare this paper as part of the United Nations International Decade for Natural Hazard Reduction.

Biodiversity and Atmospheric Change (1994) was a project undertaken to describe the linkages between the atmosphere and biological diversity and to explore the possible effects of climatic change on biodiversity. Biodiversity refers to the genetic diversity within species, the diversity of species, and the diversity of ecosystems. The project was undertaken with colleagues from Environment Canada and the University of Toronto.

Saskatchewan in a Warmer World: Preparing for the Future (1992) was a project whose goal was to improve our knowledge of the possible impacts of a warmer climate on the Saskatchewan environment and economy, including agriculture, forests, water resources, tourism and recreation, energy, and health. The prairie economy and environment are among the most climate sensitive in the world. The region is also expected to have the greatest rate of changing climate in the world. Further global warming is likely to bring an increased frequency and intensity of droughts and heat waves.

Because of our previous work (1988), we had preliminary estimates of the effects of continued climate warming on agriculture, but none were available for other sectors, such as energy or health. We worked with people from seven provincial government departments and interviewed about thirty experts for this project.

Other climatology projects run a gamut of topics. Some researchers are working on severe storms, their characteristics and hazards to people and property. Others are focussing on how weather affects insects such as grasshoppers, or diseases such as sleeping sickness.

Would You Happen to Know. . . ?

Many people call the Saskatchewan Research Council with problems and questions. The weather- and climate-related calls are routed to the Climatology Section. We answer hundreds of questions each year. After we have tracked down the solutions, we usually ask, "Why do you want to know?"

The most common requests are for temperature and precipitation information. Many people want to know the weather extremes—information needed for agricultural, horticultural, and engineering applications. Or winter temperature data may be required for studies of the effect of cold on livestock, or the winter survival of plants. Snowfall data are required to determine the snow loads that roofs must withstand.

No, it's not a crystal ball. This beautiful instrument is a vital sunshine recorder. (Virginia Wittrock)

Climatological Research Relevant to the Prairies: Recent Examples

Project	Researcher	Institute
Industrial cloud, fog and precipitation during very cold weather	R. B. Charlton	University of Alberta
Severe storm surveys	R. B. Charlton	University of Alberta
Windbreak flow	J. D. Wilson	University of Alberta
Short range atmospheric diffusion	J. D. Wilson	University of Alberta
Possible impact of climatic change on the water balance in the Brandon region of southwestern Manitoba	R. A. McGinn	Brandon University
Detection and prediction of road surface icing conditions	A. B. Shaw	Brock University
Mapping the chinook belt, signal strength and frequency	L. C. Nkemdirim	University of Calgary
Impacts of climate on seasonal water consumption in Calgary	A. Akuoko-Asibey et al.	University of Calgary
Predicting temperature and precipitation time series scenarios for Western Canada due to projected climate change	J. M. Byrne	University of Lethbridge
Agricultural response to climatic variability	Q. P. Chiotti	University of Lethbridge
Weather and insects	D. L. Johnson	University of Lethbridge
Testing the validity of historical climatic data	A. J. W. Catchpole	University of Manitoba
Prairie thunderstorms and hailstorms	A. H. Paul	University of Regina
Urban microclimate of Saskatoon	O. W. Archibold, E. Ripley	University of Saskatchewan
Hudson Bay journals as a source of climatic information	T. F. Ball	University of Winnipeg
Hailstorm hazard in the Great Plains	D. Blair	University of Winnipeg
Weather factors in the prediction of western equine encephalitis epidemics in Manitoba	R. F. Sellers, A. R. Maarouf	Atmospheric Environment Service, Environment Canada

(compiled from Rasid 1995)

Strange questions I have been asked include:

> "What was the climate during the 1200s A.D.?"—from archaeologists who were looking for pieces in the puzzle of the migration patterns of early people on the prairies. They suspected a linkage of their movements with climatic variations from decade to decade.
>
> "What was the time of sunset at Kyle, Saskatchewan, in 1894?"—from a writer who was setting the stage for an event in his book.
>
> "Were there droughts in the 1890s?"—from a researcher who was examining the conditions for development of the prairies during the late 1800s.

Other interesting and odd questions concerned the use of cloud and bright sunshine data to explain bees' activities, temperature and precipitation data as related to Saskatoon berry frost and insect damage, and temperature data and fish productivity.

Most requests come from the public: schools, libraries, media, businesses, universities, and colleagues at the Saskatchewan Research Council. The nature of the request is also reflected in the area of application. These include agriculture, recreation, law, education, media, wildlife and natural resources, environment, geology, hydrology, engineering, animal and human health, and gardening. You can usually guess the caller's background from the nature of the question.

The fact that we get all these questions is a reflection on the importance of weather and climate for human activity. Just when I think I may have a grasp of the myriad uses of climate information, a phone call with another surprising question reminds me that I'm only looking at the tip of the iceberg. Because weather and climate affect so many aspects of life, from business to beer drinking, almost any type of request can be expected.

At the Saskatchewan Research Council, the main applications of climatic information are agriculture, hydrology, forestry, engineering, and environmental impact work. A common agricultural use of climate data is the modelling of plant growth and productivity. For example, a joint study with the University of Saskatchewan and a student at SRC was to simulate the growth and yield of spring wheat for the Saskatoon area using a computer model. The model used the daily temperature, precipitation, solar radiation, and bright sunshine data from our Climate Reference Station. They "grew" wheat in the computer, simulating years of crop growth. This

Examples of Requests for Climatic Information

Discipline/Field of Application	Request	Purpose of Request
Recreation	• wind speed and direction for a specific location • temperature and precipitation normals and variations	• to facilitate planning of a marina and breakwater • information for a tourism booklet for locations in northern Saskatchewan
Law	• area, extent, and severity of a hailstorm • rates of rainfall	• evaluation of an insurance claim • evaluation of flooding insurance claims
Education/Media	• tours of the Climate Reference Station • frost season information • daily, monthly, and extremes of precipitation, temperatures, wind speed, relative humidity, and bright sunshine	• TV science shows, field information for class tours • winter wheat presentation at a conference • *Star Phoenix* readers, various reasons
Engineering	• soil temperatures • global radiation • heating degree-days	• effects on pressure in gas lines and buried cables, pipeline stress calculations • for consideration of efficiency of low energy housing • to compare with heating and ventilation costs, effectiveness of insulation
Wildlife Biology	• mean annual air temperatures • air temperature	• to evaluate fish productivity • to evaluate the caloric intake of falcons
Agriculture/ Horticulture	• temperature extremes • precipitation, growing degree-days and frost • daily temperature and precipitation	• temperature tolerances of cacti in Saskatchewan • pasture productivity • relationship to cattle health problems
Geology/ Geohydrology	• monthly precipitation • daily air temperature and wind information	• compare with ground water levels • to explore the reasons for sersonic traces and explosive noises

(after Wheaton 1987)

A climate technologist demonstrates the use of weather instruments. (Elaine Wheaton)

type of work helps improve our understanding of the relationship of weather and climatic conditions with plant growth and yield. We have also used such models to estimate the effect of future climate change on plant productivity.

Another agricultural and environmental problem that requires weather and climate information is the drifting of herbicides. When herbicide is applied, a certain proportion of the spray tends to drift from its targeted area. The SRC has worked with Agriculture Canada to evaluate and compare the performance of different types of sprayers under different weather conditions. Wind-speed data for different heights above the ground are important for these applications, for wind speed affects the amount, distance, and direction of drift.

From Calgary to Kalamazoo

I have often been asked to speak at schools for occasions such as career days, and at science institutes. I have collected a set of career opportunity examples for my talks. They give a flavour of what activities the jobs may entail, including designing and managing climatic databases, answering requests for climatic information, managing natural resources, initiating and supporting research on the effects of climate, doing field work, teaching, gaining expertise in Chinese studies, even speaking at schools.

Career opportunities for climatologists range from Calgary to Kalamazoo, from Boulder to Hong Kong. Even based in Saskatchewan, I have been fortunate enough to go to Brazil in January and subtropical China in November. I have also, and less fortunately, been invited to Guadalajara in August and Edmonton in January.

A popular question from students is, "How much money do you make?" I am used to that one. We earn about the same as engineers, but much less than basketball players and some computer magnates.

Another common question is, "Do you have kids?" Now, I know that women scientists are uncommon, but some students seem to be unsure if we are physically capable of bearing children. I soon dispel this concern by telling them about my twin sons.

Another question is, "How do you become a climatologist?" First, you should have a passionate interest in the weather and its effects. Then you need to take university training in physical geography, specializing in climatology. Climatology requires expertise in mathematics, computers, and physics, so your high-school interest and background in these subjects should be solid. Other subjects in physical geography that mesh with climatology include hydrology, geomorphology, regional geography, geographic information systems, and remote sensing. Courses in biology and agriculture are also useful, and you may be interested in economic geography because of the economic effects of certain climatic hazards such as droughts. A Master's degree or a doctorate, or their equivalents, are often listed as a requirement in career advertisements.

I stumbled onto my first employers and subsequent mentors through a computer library search. By the time I had entered my second year of university, I was tired of calculus, statistics, and physics without applications. I wanted to study something that had some connection with nature and people. I found the connection with geography, and decided to specialize in climatology. It was while I was doing research for a paper on shelter belts and snow cover that the computer politely informed me that most of the material I wanted was at the Saskatchewan Research Council. "At the *what*?" I thought. I had never heard of the place, and was aggravated to have my research interrupted. It was a cold day, so I checked the map to find the building and made plans to go there on my next free afternoon.

The SRC was in an old building out in the "boonies" on campus. The library was in the basement, crammed with books and reports. In fact, the entire building, including the hallways, seemed to be piled with books, files, reports, computer paper, and instruments. The library was outstanding, although many of the books were tucked away in the offices of the scientists. The librarians were also helpful and treated me like a real person, not just a pesky student. I was soon asked to sit in on coffee sessions with some of the scientists. I was pleased that they were so friendly. There were no women scientists at the SRC in those days—there are still only a few—but I took them up on their coffee offers and listened to them inventing their way through the day.

The next spring, a climatologist named John Bergsteinsson (Bergy to most people) hired me as a summer student at the SRC. It was a brilliant opportunity. I learned how to take weather observations and set up instruments. I also climbed down a tunnel under an enormous lysimeter (a buried tank filled with soil to measure soil moisture), and ran the big computer on campus.

One of the highlights of that summer was learning what other scientists were working on. One fellow asked us to help him differentiate smells in the materials he was working with. One smelled like bananas, another like perfume, another like something else. The last was unmistakably manure. His project was to find a way of dealing with the odour of manure. But he had lost his own sense of smell and had to entice naive summer students to help him.

Since then, my career has taken me into an area in which I am often the odd person out. Earlier, I was usually the only woman at meetings, conferences, coffee breaks, field trips. At one conference in the early 1980s, the chairperson greeted us with, "Good morning, Elaine and Gentlemen." One good thing that came of this was that I usually had a washroom all to myself, while twenty to fifty men had to share theirs. The balance is slowly changing. We now have mentor programs for female students in the sciences, so they will have role models and support, and be aware of the career possibilities for women in the sciences.

The Climatology Section in the 1990s

The Climatology Section at the Saskatchewan Research Council is composed of two research scientists, myself and Virginia Wittrock, and a technologist, Carol Beaulieu. We are in the Environment Branch, which also includes the Atmospheric Sciences, Air Quality, Aquatic Ecosystems, Plant Ecology, and Geo-environment/Ground Water Sections. The SRC's expertise covers the environment from (below) the ground up into the atmosphere. Disciplines include biology, chemistry, hydrology, environmental engineering, meteorology, and climatology.

The nature of our projects often calls for the combined expertise of several branches and disciplines. In 1989, for example, we led a project to assess the environmental and economic impacts of the 1988 drought. We were ably assisted by biologists, hydrologists, hydrogeologists, and others at the SRC. Outside the SRC, our main partners were Dr. Louise Arthur, an agricultural economist at the University of Manitoba, and her graduate student, Brenda Chorney. The project was undertaken for various government departments, including arms of Environment Canada and Agriculture Canada. Their representatives also provided advice and direction for the project.

Our official mission in the Climatology Section is to develop and transfer climatic information that will lead to environmental, economic, and social benefits. For example, we carry out computer modelling experiments to determine the effects of certain climatic conditions on crops, soil erosion, and forest fire risk. We explore strategies—different farming techniques would be one strategy—to make the effects beneficial, or at least less harmful. Then we test them with computer models to find out how well they work and where they need improvement.

The Climatology Section has worked for a wide variety of clients, from students and the general public to governments and the private sector. Among many others, our clients include the International Partnerships Fund, the Government of China, the Saskatchewan Science Teachers Association, the Saskatchewan Wheat Pool, and the Esquel Group Foundation of Brazil, an international organization that promotes balanced and sustainable economic development. Our multi-partner projects bring us a wide variety of clients and colleagues, not to mention learning experiences.

On to Austria, China, Brazil. . .

I have found out more about the climate of the Canadian prairies when I have worked in other countries than I have at the SRC. My first international work was in Austria. A well-known climatologist, Dan Williams, had described to me a project he was working on and asked if I had any ideas about climate models that would be suitable. I happened to be working on a model that used climate data to describe the potential for wind erosion of soil, and I suggested that it might be suitable.

Dan was working with an English climatologist, Martin Parry, who was at the International Institute of Applied Systems Analysis in Austria. Martin was working on a project that used case studies of different countries, and involved teams of scientists from those countries. The first of the general circulation model results for future climates in a world with higher carbon dioxide (a leading greenhouse gas) was just available then. The objective of Martin's international project was to explore the implications of such climatic variations for agriculture in cold, temperate, and semi-arid regions.

Saskatchewan was chosen as a case study area for several reasons. It is an important producer and exporter of agricultural commodities, especially wheat, and it has a tough (we called it marginal) and demanding climate for agriculture. These types of locations, we found, were expected to experience the greatest rates of climate change related to the greenhouse effect.

The project leader's search pointed to me, a climatologist with expertise in agricultural climatology in Saskatchewan who had useful ideas for climate impact assessment. Those circumstances led me to the first of many international work experiences. The Canadian team included four other scientists. We worked with teams from Austria, Iceland, Finland, Russia, and Japan. Several years later we produced one of the first books on the effects of present and future climate variations on agriculture.

I have since been to Austria again, to China (twice), and to Brazil. The travel and the international work has brought me a wealth of experience, and I have learned a great deal about the intricate interconnections of our prairie climate in comparison with other areas.

References

Lazar, A. 1995. Canadian commitments to biodiversity. In *Proceedings, national meeting on the ecological monitoring and assessment network*. Burlington, ON: Ecological Monitoring Coordinating Office, Canadian Centre for Inland Waters.

Rasid, H. 1995. *The Canadian Association of Geographers directory 1994*. Thunder Bay: Department of Geography, Lakehead University.

Saskatchewan Research Council (SRC). 1995. *1994–1995 annual report*. Saskatoon: Saskatchewan Research Council.

Thomas, M. K. 1985. The value of paleoclimatic data to a climatologist. In *Climatic change in Canada: Critical periods in the Quaternary climatic history of northern North America*, edited by C. R. Harington. In *Syllogeous 55*. Ottawa: National Museums of Canada.

Wheaton, E. E. 1987. Applications of climatology: The current Saskatchewan Research Council experience. Presented at the *Alberta Climatological Association's 11th annual meeting, February 24, 1987, Edmonton, AB*. Saskatchewan Research Council (SRC) Pub. No. E-906-4-D-87. SRC, Saskatchewan.

Extreme cold caused snow and ice to accumulate on the heated boilers of this steam locomotive.
(South Saskatchewan Photo Museum)

Appendix I

What Is a Snow Roller?

Quizzes on Climate

Storm Climaquiz

(Note: answers and comments are on pages 167–169)

1. Most Canadian regions experience blizzards, but the blizzard alley is in:
   1) the coast of British Columbia
   2) the prairie provinces
   3) Newfoundland
   4) the Arctic

2. At least four of Canada's ten worst blizzards have occurred on the prairies.
   ❑ T ❑ F

3. Prairie blizzards are winter phenomena only and do not occur in the spring or summer.
   ❑ T ❑ F

4. Blizzards are localized storms with narrow paths and usually affect only small areas.
   ❑ T ❑ F

5. Blizzards cause disruption and heartache, including destruction of buildings, crops, livestock, and death.
   ❑ T ❑ F

6. Dust storms are ten to twenty times as common in the spring than at other times of the year.
   ❑ T ❑ F

7. Dust storms are never experienced in the winter on the prairies.
   ❑ T ❑ F

8. Dust storms are "kind" storms; they have few effects and have never caused any deaths on the Canadian prairies.
   ❑ T ❑ F

9. Dust storms, or black blizzards, are another name for a dust devil or whirlwind.
   ❑ T ❑ F

10. Mud storms have occurred on the prairies.
    ❑ T ❑ F

11. Tornadoes are nature's most locally destructive storms.
    ❑ T ❑ F

12. A tornado is a violently rotating column of air hanging from a cumulo-nimbus cloud.
    ❑ T ❑ F

13. A tornado's wind speeds are estimated to be:
    1) less than 100 km/h
    2) 100 to 200 km/h
    3) 100 to 300 km/h
    4) 100 to over 400 km/h
    5) none of the above

14. Which country has the greatest risk of tornadoes?
    1) Brazil
    2) Maldives
    3) Canada
    4) Siberia
    5) United States

15. What was the most damaging tornado, in terms of lives lost?
    1) Edmonton, 31 July 1987
    2) Gimli, 3 June 1946
    3) Saskatoon, 1 June 1986
    4) Regina, 30 June 1912
    5) Calgary, 5 August 1954

16. Tornadoes never strike the same place twice.
    ❑ T ❑ F

17. Even the weakest tornadoes can push over trees, move empty granaries, break signs, and destroy chimneys.
    ❑ T ❑ F

18. A tornado is identified and classified by which of the following criteria:
    1) a funnel cloud touching the ground
    2) an explosion
    3) debris arranged to indicate the passage of a vortex
    4) dense, heavy objects lifted
    5) roofs taken off buildings along a narrow, extended path
    6) all of the above

19. The more severe the tornado, the longer the path.
    ❑ T ❑ F

20. Prairie tornadoes are most common in:
    1) April to September
    2) May to June
    3) late June to early July
    4) early August to late September
    5) no set time; occurrences are random

21. Prairie tornadoes usually originate out of the south, southwest, or west.
    ❑ T ❑ F

22. Tornadoes have wide, predictable paths.
    ❑ T ❑ F

23. Tornadoes are rarely accompanied by hail and lightning.
    ❑ T ❑ F

24. Most mobile homes can easily survive a tornado.
    ❑ T ❑ F

25. Tornadoes have been known to:
    1) cause frog showers
    2) de-feather birds
    3) remove horse hair
    4) bend and rip railroad ties
    5) drive straw into steel pipes
    6) all of the above

PHOTO 6-1: What's going on? (Karen Wheaton)

26. A tornado-wise person will:
    1) stay away from windows, doors and exterior walls
    2) go into the basement and seek shelter under the stairs or a sturdy table
    3) avoid mobile homes; if you are in one, get out and seek shelter in a sturdy building, ditch, or culvert
    4) avoid wide-span buildings such as barns, auditoriums, and supermarkets
    5) go to lower levels in high-rise buildings and find shelter in small interior rooms or stairwells
    6) all of the above

27. Tornadoes sound like:
    1) the rush of a thousand freight trains speeding through a tunnel
    2) the clattering of thousands of Venetian blinds
    3) a giant blow-torch

4) the booming of thousands of artillery shells
5) the roar of a million bees
6) any of the above

Seasons Climaquiz

1. During winter, air masses bring topsy-turvy weather to the prairies. Daily high temperatures can be recorded at night instead of during the day. Southern stations can be much colder than northern stations.
   ❑ T ❑ F

2. Seasons are defined:
   1) astronomically
   2) according to temperature
   3) by month groupings
   4) according to precipitation
   5) all of the above
   6) none of the above

3. The prairie grassland region is:
   1) the driest in Canada
   2) the driest in the world
   3) the second driest in Canada
   4) the second wettest in Canada
   5) none of the above

PHOTO 6-2: What is it? (Elaine Wheaton)

PHOTO 6-3: What is it? (Elaine Wheaton)

4. Vancouver has more pronounced temperature seasons than Winnipeg.
   ❑ T ❑ F

5. The sunniest region in Canada is:
   1) coastal British Columbia
   2) the Arctic
   3) the southern prairies
   4) none of the above

6. Average temperatures decrease much more rapidly northward in the summer than in the winter on the prairies.
   ❑ T ❑ F

7. Isotherms are lines joining points:
   1) of equal elevation
   2) of equal meaning
   3) of different precipitation
   4) of equal intelligence
   5) of equal temperature
   6) none of the above

8. The prairies hold the record for both the hottest and the coldest temperatures.
   ❑ T ❑ F

9. Which major prairie city has the lowest January average temperature?
   1) Calgary
   2) Lost River
   3) Winnipeg
   4) Saskatoon
   5) none of the above

10. Regina is the major prairie city with the lowest average January wind-chill; therefore it feels the coldest.
    ❑ T ❑ F

11. Which major cities hold the record low temperature for the prairies?
    1) Winnipeg
    2) Regina
    3) Calgary
    4) Edmonton
    5) Saskatoon

12. The lowest wind-chill in Canada was recorded at:
    1) Regina
    2) Edmonton
    3) Tuktoyaktuk
    4) Pelly Bay
    5) none of the above

13. The record low temperature for the prairies was recorded at:
    1) Churchill
    2) Snag
    3) Norway House
    4) Fort Vermilion
    5) Prince Albert
    6) Kilmahumaig

14. The chance of a "bonspiel thaw" in midwinter decreases from Alberta eastward to Manitoba.
    ❑ T ❑ F

15. Wind-chill makes you feel cold because:
    1) you have to work harder to stand upright
    2) you sweat more rapidly
    3) your body heat is removed more rapidly
    4) your clothes are blown against your body
    5) none of the above

16. The temperature of an object will be reduced to below the surrounding air temperature during a severe wind-chill.
    ❑ T ❑ F

17. Less damage is suffered if a person, animal, or plant freezes rapidly.
    ❑ T ❑ F

18. Cold weather kills more often than any other type of weather, including heat waves, lightning, tornadoes, and floods.
    ❑ T ❑ F

19. Which province has the record greatest depth of snow cover?
    1) British Columbia
    2) Alberta
    3) Saskatchewan
    4) Manitoba
    5) Quebec
    6) none of the above

20. Prairie snow cover has been blamed for affecting the results of political nominations.
❑ T ❑ F

21. A snow roller is a rare phenomenon on the prairies. What is it?
1) the rolled crest of a snow bank
2) a fluffy white type of hair curler
3) the cylindrical portion of a hail stone
4) a cylindrical snowball caused by the wind
5) it does not exist

22. The last frost in spring is expected about mid-April on the southern prairies.
❑ T ❑ F

23. February and August are usually the coldest and hottest months on the prairies.
❑ T ❑ F

24. Heat waves are hazardous to humans, animals, and plants.
❑ T ❑ F

25. The Canadian record for high temperatures is held by:
1) Manyberries, Alberta
2) Yellowgrass, Saskatchewan
3) Waskada, Manitoba
4) Rimouski, Quebec
5) none of the above

26. The summer of 1816 is known for its record warmth in North America and the world.
❑ T ❑ F

PHOTO 6-4: What are they? (Rey Marr)

PHOTO 6-5: What is this? (Elaine Wheaton)

27. Winter is the wettest season on the prairies.
    ❑ T ❑ F

28. The southern prairies average how many hot days (temperature greater than 20°C) per year?
    1) less than 10
    2) less than 5
    3) more than 20
    4) more than 40

29. Where did Canada's record hail stone fall?
    1) Montreal, Quebec
    2) Ottawa, Ontario
    3) Yellowknife, North West Territories
    4) Cedoux, Saskatchewan
    5) none of the above

30. Hail storms usually occur during the hottest part of the year.
    ❑ T ❑ F

31. Crop insurance payments for hail damage in Saskatchewan are greater than for drought in most years.
    ❑ T ❑ F

32. The urban hail hazard has been decreasing through time.
    ❑ T ❑ F

33. The Canadian hail storm alley extends through Prince Albert, Saskatchewan.
    ❑ T ❑ F

34. Spring is usually warmer than fall.
❑ T ❑ F

35. Fall frosts usually strike during mid-September for most of the agricultural prairies.
❑ T ❑ F

36. Spring and fall frosts are dependable from year to year and place to place (i.e., frosts do not depend on such factors as distance from lakes or the slope of the ground).
❑ T ❑ F

37. Temperatures of about 10°C feel warmer in the spring than the fall because humans acclimatize.
❑ T ❑ F

38. Summer weather is dominated by the sun's control and winter weather is dominated by air mass control.
❑ T ❑ F

39. The prairie climate is predictable, has few storms, and is not very exciting.
❑ T ❑ F

Drought Climaquiz

1. Droughts are most intense, common, and widespread:
   1) in the interior of British Columbia
   2) in the Arctic
   3) on the prairies
   4) in southern Quebec
   5) none of the above

2. Droughts are not considered major natural disasters.
   ❑ T ❑ F

3. Drought can be measured by:
   1) the number of famines
   2) the number of wells that go dry
   3) its effect on crop yields
   4) its effect on wildlife
   5) its effect on the brewing industry
   6) all of the above
   7) none of the above

4. Areas with lowest rainfall tend to have more reliable rainfall.
   ❑ T ❑ F

5. Prairie rainfall is sporadic from place to place and year to year.
   ❑ T ❑ F

6. Droughts can affect:
   1) water quality
   2) soil erosion
   3) air quality
   4) human and animal health
   5) insects
   6) all of the above

7. Patterns of sea surface temperatures in the Pacific do not affect the prairies' climates.
   ❑ T ❑ F

8. The boreal forest on the northern prairies is not affected by drought.
   ❑ T ❑ F

9. Droughts affecting prairie crop yields occurred in:
   1) 1910
   2) 1924
   3) 1941
   4) 1958
   5) 1974
   6) 1990
   7) all of the above

10. The most severe and devastating prairie droughts occurred in the 1930s and 1980s.
   ❑ T ❑ F

11. Saskatchewan was the prairie province hardest hit by the 1980s drought.
   ❑ T ❑ F

12. Weeds tend to do poorly during droughts, and few weed management problems occur.
   ❑ T ❑ F

13. Socioeconomic conditions were worse in the 1980s as compared to the 1930s.
   ❑ T ❑ F

14. The 1930s drought resulted in unprecedented farm abandonment.
   ❑ T ❑ F

15. The risk of future drought on the prairies is low.
   ❑ T ❑ F

16. Future droughts are expected to increase in frequency and intensity because of continued global warming.
   ❑ T ❑ F

17. A commonly used livestock feed and building material during the 1930s was a weed known as the Russian thistle.
   ❑ T ❑ F

Future Climaquiz

1. Greenhouse gases are atmospheric gases that absorb the earth's heat. They include:
   1) carbon dioxide
   2) water vapour
   3) chlorofluorocarbons
   4) methane
   5) nitrous oxides
   6) all of the above
   7) none of the above

2. Human activities that release greenhouse gases include:
   1) driving cars
   2) using electricity generated from coal-fired power plants
   3) using landfills, or garbage dumps
   4) using fertilizers
   5) growing rice
   6) burning forests
   7) all of the above

3. The current level of carbon dioxide in the atmosphere is greater than the highest values of the past 160,000 years.
   ❑ T ❑ F

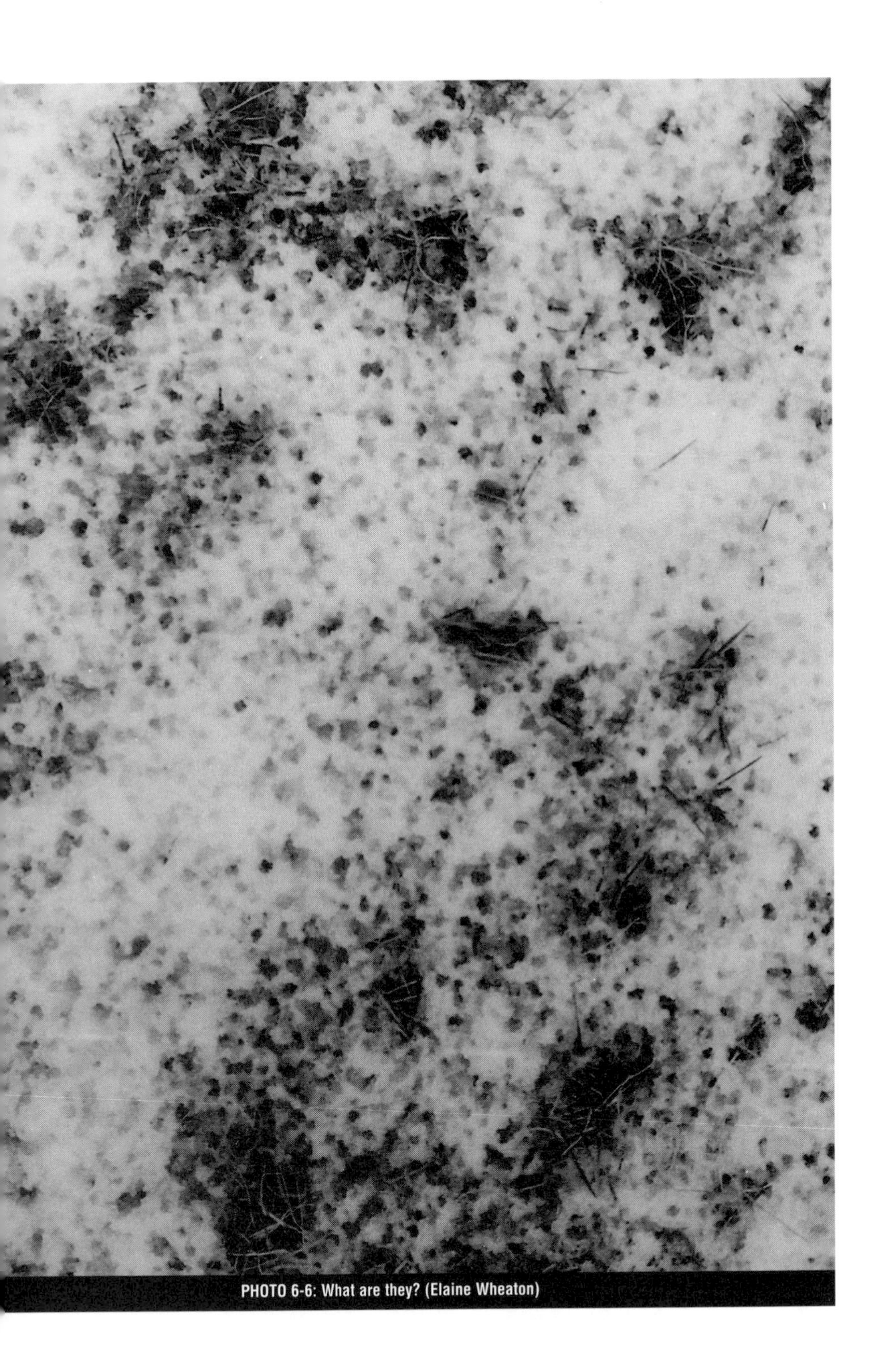

PHOTO 6-6: What are they? (Elaine Wheaton)

4. How much carbon from human activities is added to the atmosphere each year?
   1) 6 grams
   2) 6 million tonnes
   3) 6 billion tonnes
   4) 60 billion tonnes

5. Atmospheric changes in terms of greenhouse gases are localized, and have not been well monitored or measured.
   ❑ T ❑ F

6. Most greenhouse gases are increasing each year.
   ❑ T ❑ F

7. Methane is a more powerful greenhouse gas than carbon dioxide, in terms of the heat absorbing capability of each molecule.
   ❑ T ❑ F

8. Carbon dioxide is the most important greenhouse gas because of its relative abundance in the atmosphere.
   ❑ T ❑ F

9. What is the greenhouse effect?
   1) the heating of the earth by greenhouses
   2) the effect of a greenhouse on a plant's growth
   3) the effect of the moon's atmosphere
   4) the trapping of the earth's heat by greenhouse gases
   5) the action of greenhouse gases to reflect the sun's energy
   6) none of the above

10. Clouds can either cool or heat the earth's surface, depending on their height above the ground.
    ❑ T ❑ F

11. Uncertainties in the projections of climatic warming include the effects of:
    1) clouds, which influence the amount of change
    2) oceans, which affect the timing of climatic change
    3) atmospheric particles, which act to cool the earth
    4) polar ice sheets, which affect the amount of change
    5) none of the above
    6) all of the above

12. The earth is about 33°C warmer than it would be without the natural greenhouse gas effect.
    ❑ T ❑ F

13. What are the causes of climatic change?
    1) changes in the earth's orbit
    2) changes in amounts of solar radiation
    3) changes in the greenhouse gases
    4) changes in the amounts of dust in the atmosphere
    5) all of the above
    6) none of the above

14. A panel of many scientists around the world has stated that human impacts are now discernible in the climate record, that past climatic changes have not been caused entirely by natural vari-

PHOTO 6-7: What are they? (Elaine Wheaton)

ability, and that future global warming of unprecedented amounts is expected.
❑ T ❑ F

15. The greenhouse effect theory is controversial.
❑ T ❑ F

16. There is considerable evidence that global warming and other changes are linked to the greenhouse effect.
❑ T ❑ F

17. What types of evidence link global warming and the greenhouse effect?
1) results from glacial ice cores
2) results from studies of other planets
3) mathematical experiments
4) trends in climatic data
5) all of the above
6) none of the above

18. Results of the models of the earth's climate system under increasing amounts of greenhouse gases show:
1) increased warming in the tropics
2) increased warming in summer
3) decreased global precipitation
4) increased precipitation in semi-arid regions
5) increased warming during the day more than the night
6) all of the above
7) none of the above

PHOTO 6-8: What is it? (John Perret)

19. The greenhouse effect is operating:
    1) to make a car very hot on a warm sunny day
    2) on a cloudy night to keep temperatures higher than on a clear night
    3) on the planet Venus
    4) to keep Earth warmer than it would be without an atmosphere
    5) all of the above
    6) none of the above

20. Reports indicate that, with increasing greenhouse gases, the earth's temperature could increase at a greater rate than any experienced in the past 10,000 years.
    ❑ T ❑ F

21. There are many actions individuals can take to decrease emissions of greenhouse gases.
    ❑ T ❑ F

22. The earth is currently in a long-term cooling trend.
    ❑ T ❑ F

23. The next ice age is estimated to return within the next 100 years.
    ❑ T ❑ F

24. The prairie region has already experienced an increase in temperatures over the past 100 years.
    ❑ T ❑ F

PHOTO 6-9: What is this? (John Perret)

25. Some of the possible effects of continued warming on the prairies include:
    1) increased intensity and frequency of droughts
    2) problems with water quantity and quality
    3) increased frequency and severity of tornadoes and dust storms
    4) increased soil erosion by wind
    5) increases in forest fires
    6) northward shifts in climatic conditions suited to plants and animals
    7) all of the above
    8) none of the above

26. Scientists know that continued warming in the prairie region will be beneficial to farming.
    ❑ T ❑ F

27. Variation in the sun's energy is the main reason that climates change. Therefore, we should not be concerned about the increasing greenhouse effect.
    ❑ T ❑ F

28. The number of prairie heat waves will increase and cold spells will decrease with continued global warming.
    ❑ T ❑ F

29. Canada's climate has cooled and dried over the past 100 years.
    ❑ T ❑ F

30. Rising sea levels are also a concern related to global warming and the melting of the polar ice caps.
    ❑ T ❑ F

31. Chlorofluorocarbons are greenhouse gases that also cause depletion of the ozone layer.
    ❑ T ❑ F

32. Only huge corporations and nations can tackle the increasing greenhouse gas problem; it is beyond the control of individuals.
    ❑ T ❑ F

Answers and Comments

Storm Climaquiz answers

1. 2 the prairie provinces
2. T
3. F
4. F
5. T
6. T
7. F
8. F
9. F
10. T
11. T
12. T
13. 4 100 to over 400 km/h
14. 5 United States (Canada has the second greatest risk of tornadoes in the world)
15. 4 Regina, 30 June 1912
16. F
17. T
18. 6 all of the above
19. T
20. 3 late June to early July
21. T
22. F
23. F
24. F
25. 6 all of the above
26. 6 all of the above
27. 6 any of the above

Seasons Climaquiz answers

1. T
2. 5 all of the above
3. 3 the second driest in Canada; the Arctic region receives the least amount of annual average precipitation in Canada
4. F
5. 3 the southern prairies
6. F
7. 5 of equal temperature
8. F
9. 3 Winnipeg
10. T
11. 2 & 5: both Regina and Saskatoon have the lowest record low at -50.0°C
12. 4 Pelly Bay
13. 4 Fort Vermilion (Fort Vermilion's record low is -61.1°C, 11 Jan. 1911; Snag, Yukon, holds the Canadian record at -63.0°C, 3 Feb. 1947; the world record is held by Vostok, Antarctica at -89.6°C, 21 July 1983)
14. T
15. 3 your body heat is removed more rapidly
16. F objects will not cool below the air temperature; wind-chill increases the rate the body is cooled to the air temperature
17. T
18. T
19. 1 British Columbia (Hudson Bay, Saskatchewan, holds the prairie record)
20. T
21. 4 a cylindrical snowball caused by the wind
22. F the average last spring frost is about mid-May on the southern prairies
23. F January and July
24. T
25. 2 Yellowgrass, Saskatchewan (Yellowgrass and Midale, Saskatchewan, hold Canada's record temperatures, 45°C, 5 July 1937)

26. F the summer of 1816 is known as the "year without a summer," and is renowned for its unusual cold
27. F summer is the wettest season; June is the wettest month in the southern prairies, July the wettest in the north
28. 3 more than 20
29. 4 Cedoux, Saskatchewan
30. T
31. T
32. F
33. F
34. F
35. T
36. F
37. T
38. T
39. F

Drought Climaquiz answers

1. 3 on the prairies
2. F
3. 6 all of the above
4. F
5. T
6. 6 all of the above
7. F
8. F
9. 7 all of the above
10. T
11. T
12. F
13. F
14. T
15. F
16. T
17. T

Future Climaquiz answers

1. 6 all of the above
2. 7 all of the above
3. T
4. 3 6 billion tonnes
5. F
6. T
7. T
8. T
9. 4 the trapping of the earth's heat by greenhouse gases
10. T
11. 6 all of the above
12. T
13. 5 all of the above
14. T
15. F the greenhouse effect theory, that the earth-atmosphere system is heated by greenhouse gases, is one of the oldest and best established theories in atmospheric science; what is controversial is how and when the increasing greenhouse gases will change the earth's climate, and what the effects may be
16. T
17. 5 all of the above
18. 7 none of the above
19. 5 all of the above
20. T
21. T
22. F
23. F
24. T
25. 7 all of the above
26. F
27. F
28. T
29. F
30. T
31. T
32. F

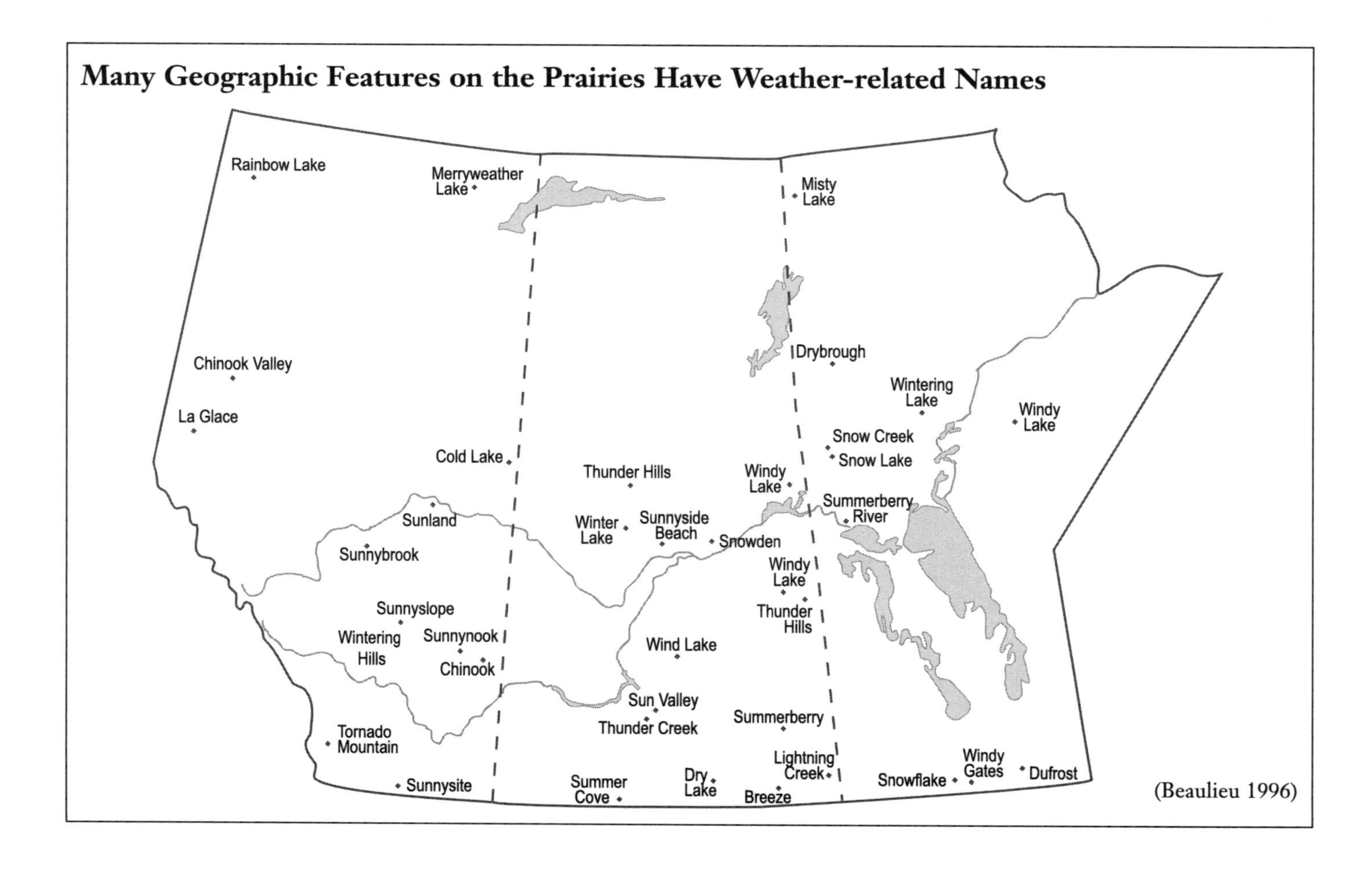
Many Geographic Features on the Prairies Have Weather-related Names
Rainbow Lake
Merryweather Lake
Misty Lake
Chinook Valley
La Glace
Drybrough
Wintering Lake
Windy Lake
Snow Creek
Snow Lake
Cold Lake
Thunder Hills
Windy Lake
Summerberry River
Sunland
Winter Lake
Sunnyside Beach
Snowden
Sunnybrook
Windy Lake
Thunder Hills
Sunnyslope
Wintering Hills
Sunnynook
Chinook
Wind Lake
Sun Valley
Thunder Creek
Summerberry
Tornado Mountain
Lightning Creek
Windy Gates
Dufrost
Snowflake
Sunnysite
Summer Cove
Dry Lake
Breeze
(Beaulieu 1996)

Photo Climaquiz

PHOTO 6-1: The author investigates a sinkhole formed by the severe water erosion during the rainy summer of 1991.

PHOTO 6-2: Snow crystals gathered by a horse's chesnut coat—an ideal way to study the crystals.

PHOTO 6-3: Soil layers sandwiched in a snowbank. Layers of wind-blown soil are deposited between snowstorms and can be used to track the number of such storms.

PHOTO 6-4: Snow rollers are cylindrical snowballs caused by the wind.

PHOTO 6-5: The hole in this rock is caused by layers peeling from the rock as a result of freeze-thaw action.

PHOTO 6-6: These ice craters are caused by raindrops falling on snow cover and melting tiny holes—Swiss cheese snow!

PHOTO 6-7: Minicraters (about 1.5 cm in diameter) in the soil caused by the impact of hailstones. Aren't you glad they didn't hit you on the head?

PHOTO 6-8: This is a hail-damaged canola crop.

PHOTO 6-9: Icy black snowbanks are a result of wind-blown soil deposited on snowbanks and polished into surface ice layers.

References

Beaulieu, C. 1996. Personal Communication. Carol Beaulieu is a Research Technologist, Climatology Section, Saskatchewan Research Council, Saskatoon.

Appendix II

Everything You Always Wanted to Know about Prairie Weather But Were Afraid to Ask

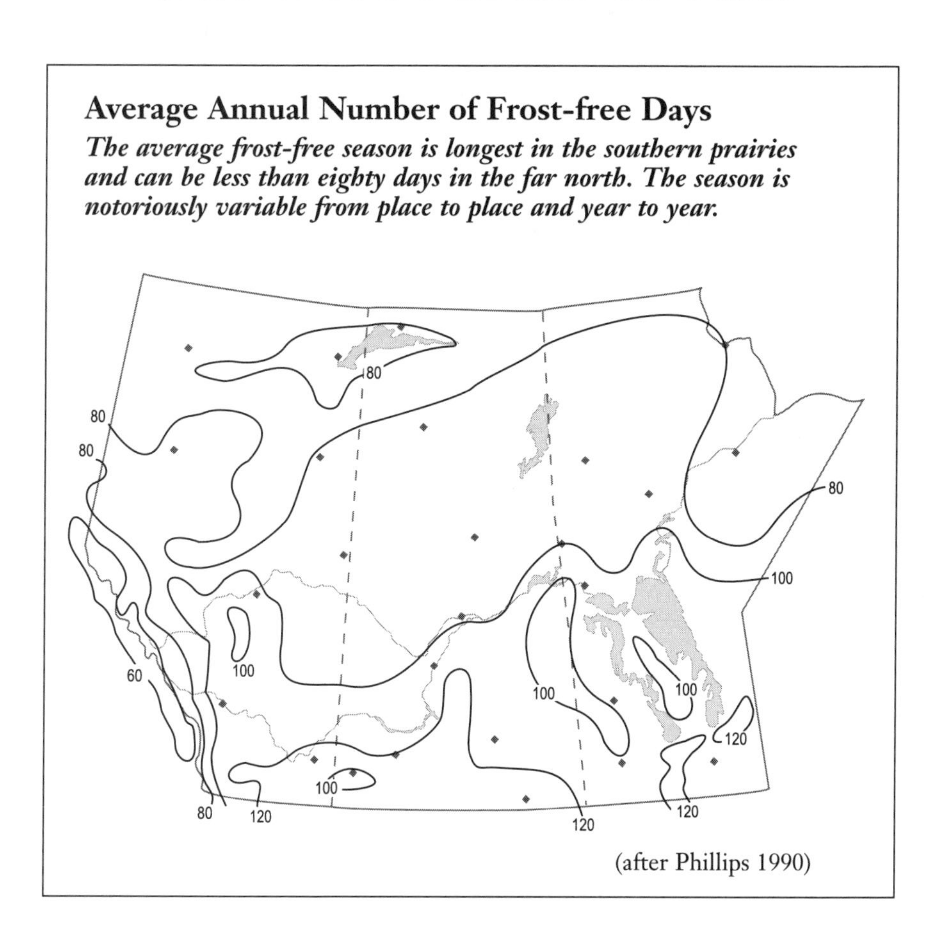

Average Annual Number of Frost-free Days

The average frost-free season is longest in the southern prairies and can be less than eighty days in the far north. The season is notoriously variable from place to place and year to year.

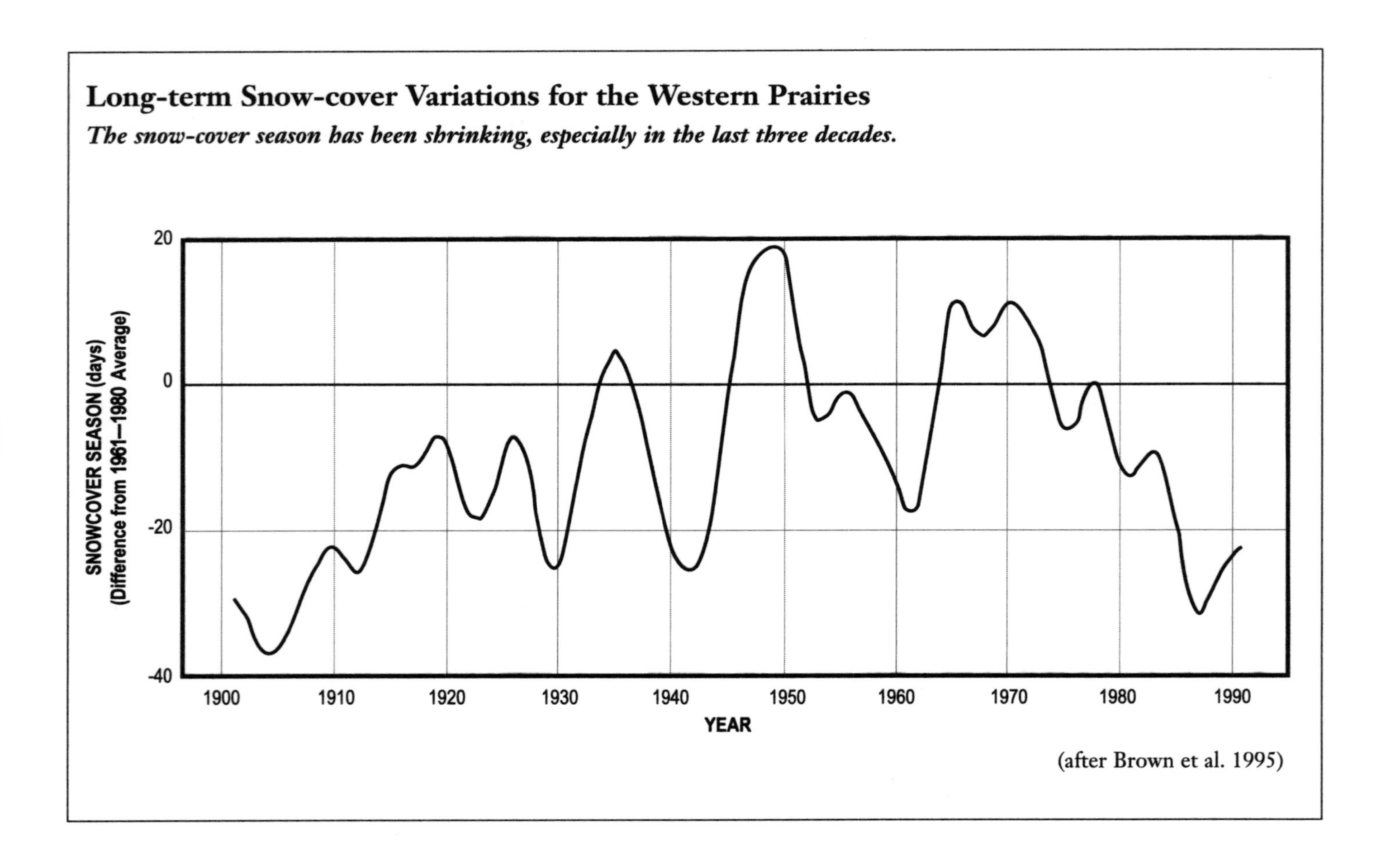

Long-term Snow-cover Variations for the Western Prairies

The snow-cover season has been shrinking, especially in the last three decades.

A spring ritual—a young boy does a balancing act after getting his boot stuck in the mud. (Gord Waldner)

Winter Air Masses and Atmospheric Circulation

Air masses clashing over the prairies can cause severe winter storms.

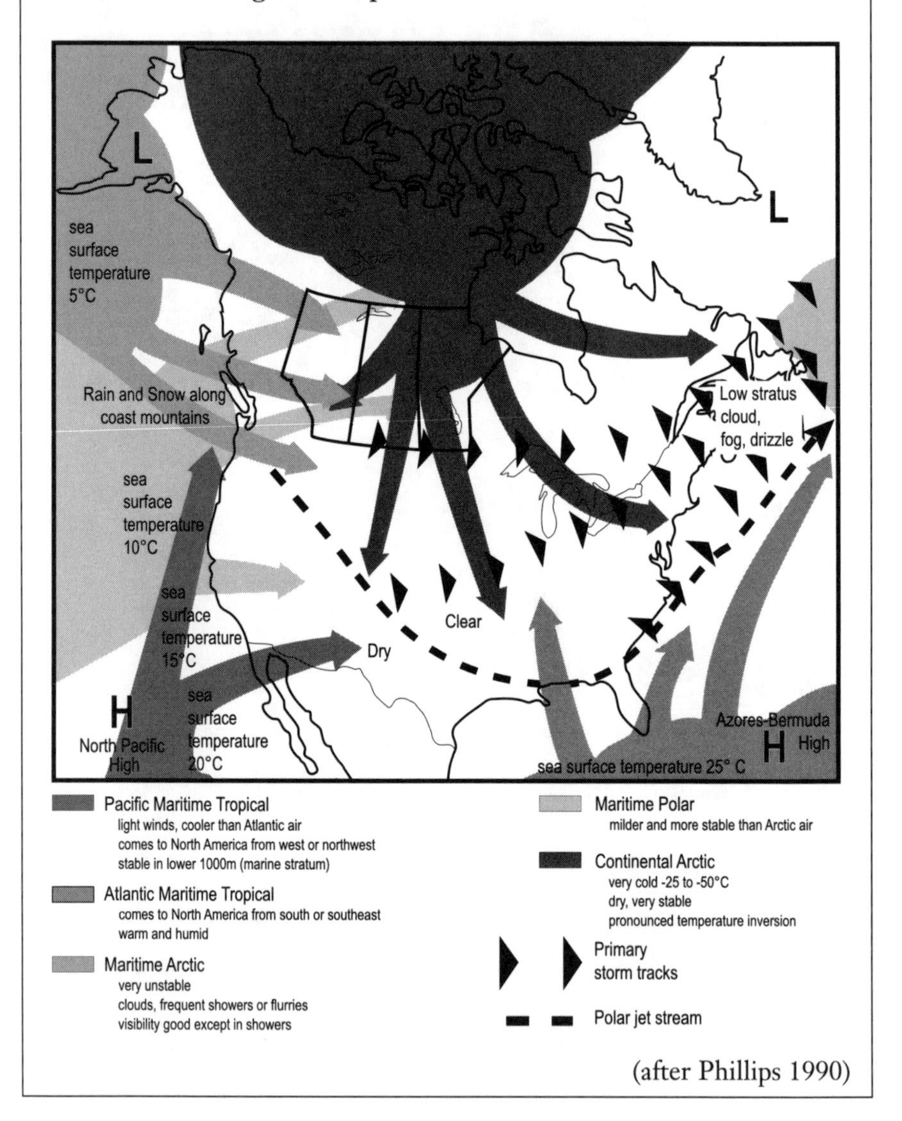

(after Phillips 1990)

Summer Air Masses and Atmospheric Circulation

Air masses from distant southern areas visit the prairies in the summer.

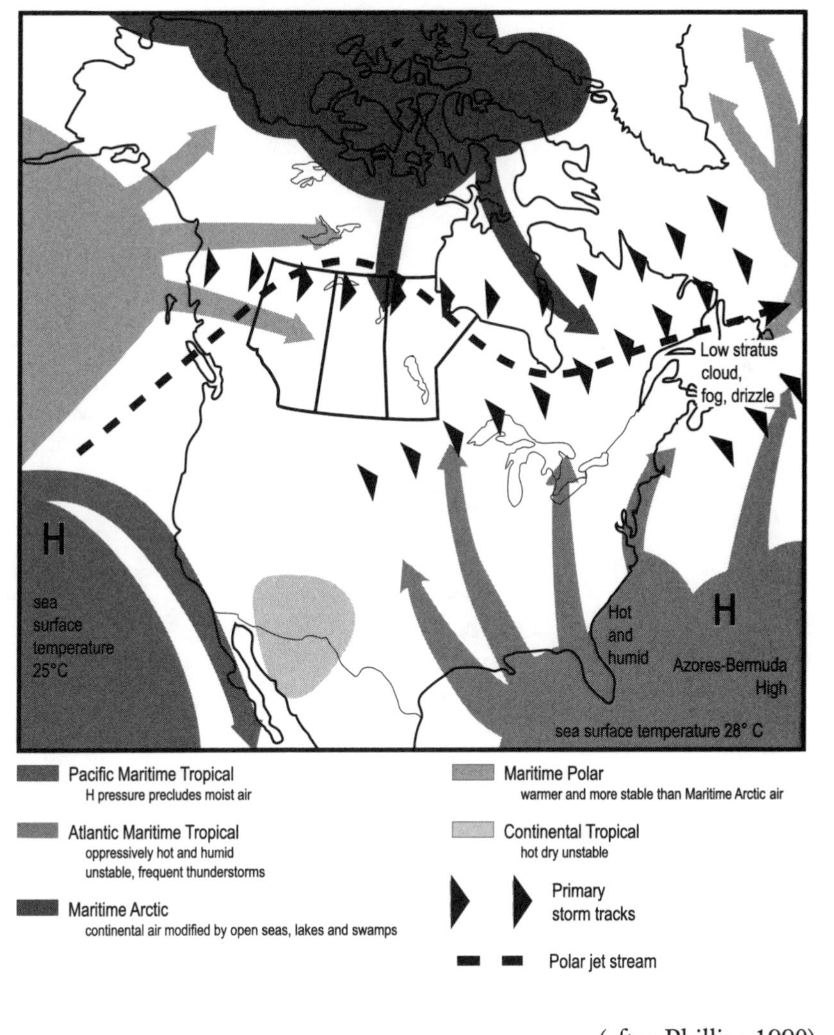

(after Phillips 1990)

The Prairie Old-timers

Most prairie climate stations are relatively recent, but here are several old-timers with records that are ninety years and longer.

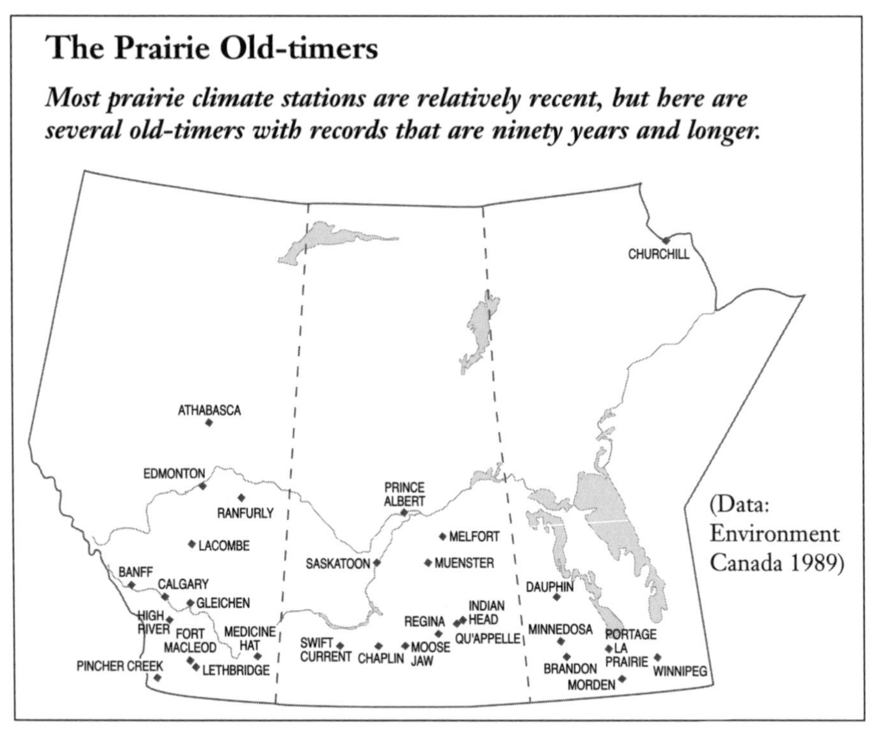

(Data: Environment Canada 1989)

Lowest Recorded Temperatures (°C) and Highest Recorded Temperatures (°C)

Prairie temperature records are truly extreme.

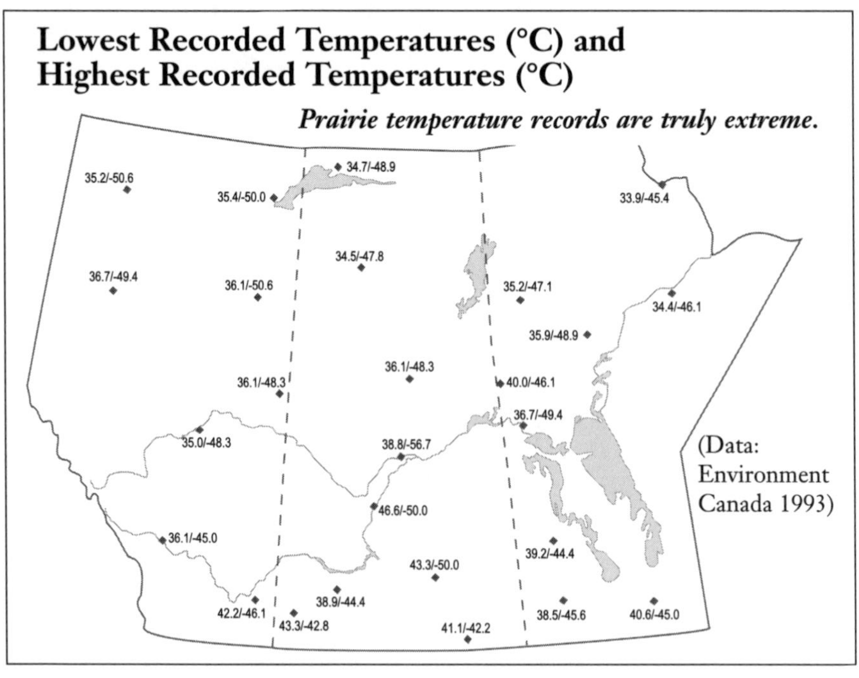

(Data: Environment Canada 1993)

Average Maximum Snow Depth (cm)

Prairie snow cover is usually lightest in the south-central region and deepest in the foothills.

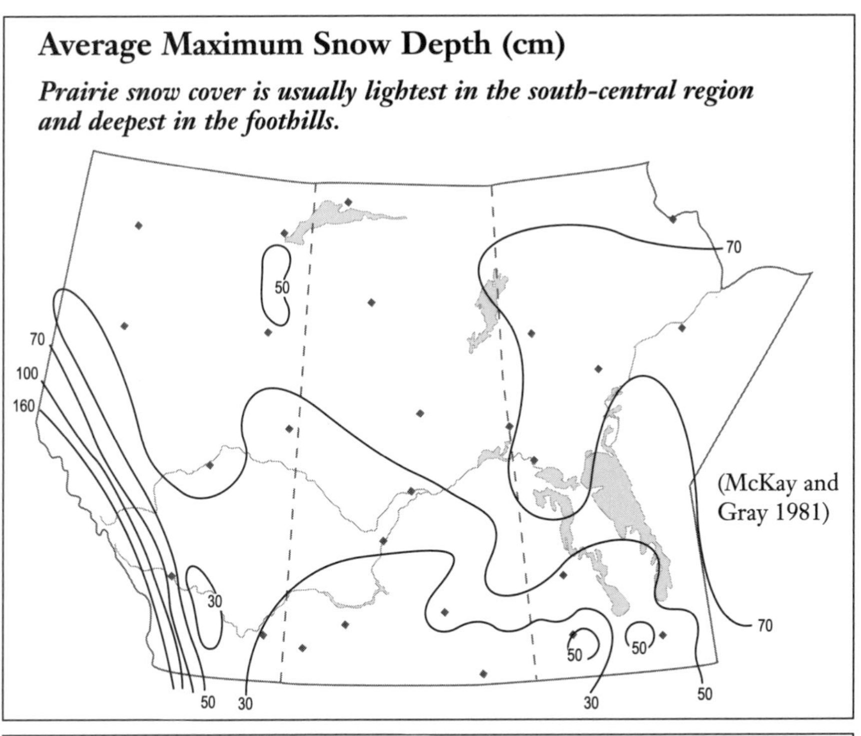

(McKay and Gray 1981)

Factoring the Wind-chill

Check this wind-chill chart before venturing outside on a harsh winter day.

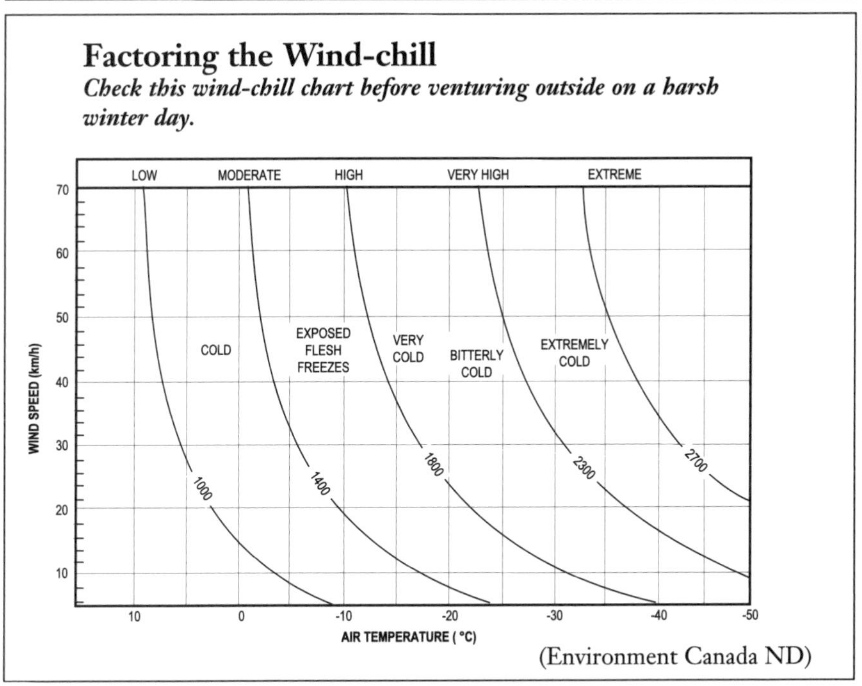

(Environment Canada ND)

July Average Daily Wind Speed (km/h) and Direction

The strongest July wind speeds occur most often in the southern prairies.

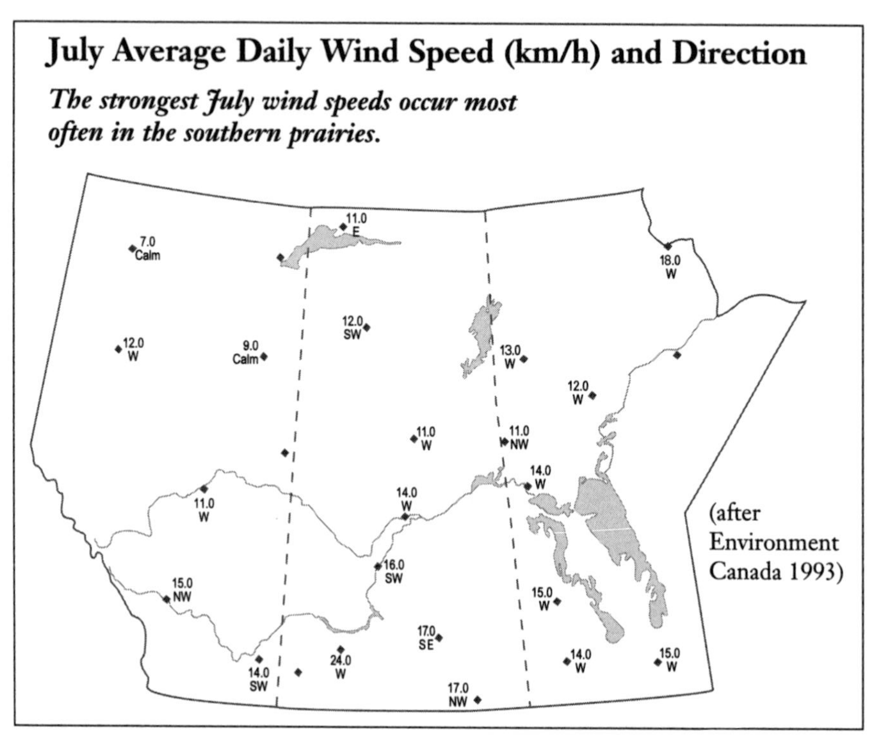

(after Environment Canada 1993)

January Average Daily Wind Speed (km/h) and Direction *Northern Alberta typically has the lowest wind speeds.*

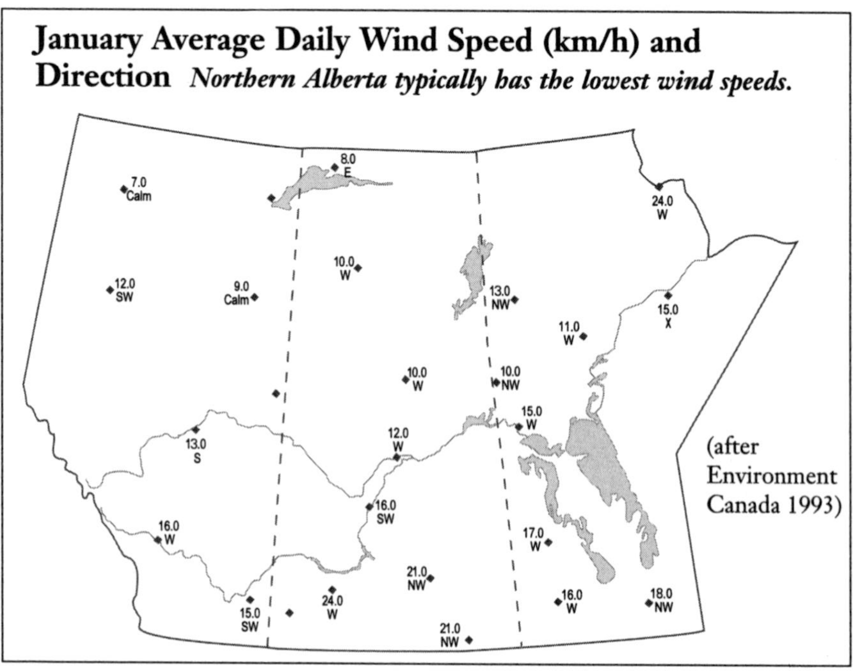

(after Environment Canada 1993)

References

Brown, R.D., M.G. Hughes, and D.A. Robinson, 1995. Characterizing the long-term viability of snow cover extent over the interior of North America. *Annals of Glaciology* 21:45–50.

Environment Canada. 1993. *Canadian climate normals on diskette*. Version 2.0E. Downsview, ON: Environment Canada.

______. n. d. *Wind-chill.* Internal Fact Sheet. Saskatoon: Saskatchewan Environmental Services Centre, Environment Canada.

Environment Canada, Atmospheric Environment Service. 1989. *Climatological station catalogue, prairie provinces*. Downsview, ON: Environment Canada, Atmospheric Environment Service.

McKay, G. A., and D. M. Gray. 1981. In Gray, D. M. and Male, D. H., eds, *Handbook of snow: principles, processes, management and use*. Toronto, ON: Pergamon.

Phillips, D. 1990. *The climates of Canada*. Ottawa: Canadian Government Publishing Centre.

INDEX

Note: *(f)* indicates a figure or table; *(p)* indicates a photograph.

About the Author

Elaine Wheaton (*StarPhoenix* Photo)

Raised on a farm near Landis, Saskatchewan, Elaine Wheaton learned first-hand to appreciate the climate of the Canadian prairies in all its rage and beauty.

Elaine has been with the Saskatchewan Research Council in Saskatoon since 1980, as well as periods during the 1970s. She is currently the Lead Scientist at the SRC's Climatology Section, where she studies the patterns, extremes, and dynamics of climate in time and space. She earned a Master of Science in Climatology in 1979, and has been an adjunct professor at the University of Saskatchewan and a member of the Faculty of Graduate College since 1992.

Elaine has written and co-authored well over a hundred scientific journal articles, technical reports, and other papers, and has given many invited presentations in Canada, the United States, Austria, Brazil, and China. She is a retired member of the Board of Governors of the Royal Canadian Geographical Society, and a fellow of the Royal Meteorological Society and Royal Canadian Geographical Society. In 1997 she received the YWCA's Science and Technology Women of Distinction Award.

Elaine lives on a farm near Saskatoon, Saskatchewan, with her husband, Dale Young, and their twin teenagers, Sean and Trent.